RODNEY A. BROOKS

Flesh and Machines

Rodney A. Brooks is Fujitsu Professor of Computer Science and Engineering at MIT and director of the Artificial Intelligence Laboratory. He is also chairman and chief technological officer of iRobot Corporation. He is a founding fellow of the American Association for Artificial Intelligence (AAA) and a fellow of the American Association for the Advancement of Science (AAAS). The author of several books and a contributor to many journals, he was one of the subjects of Errol Morris's 1997 documentary, *Fast, Cheap, and Out of Control*. Brooks was born in Australia and now lives in suburban Boston.

Also by Rodney A. Brooks

Model-Based Computer Vision

Programming in Common Lisp

Artificial Life IV: Proceedings of the Fourth International Workshop on the Synthesis and Simulation of Living Systems (co-edited with Pattie Maes)

The Artificial Life Route to Artificial Intelligence: Building Embodied Situated Agents (co-edited with Luc Steels)

Cambrian Intelligence

Flesh and Machines

Flesh and Machines: How Robots Will Change Us

Rodney A. Brooks

Vintage Books   A Division of Random House Inc.   New York

FIRST VINTAGE BOOKS EDITION, FEBRUARY 2003

Copyright © 2002 by Rodney A. Brooks

All rights reserved under International and Pan-American Copyright Conventions.
Published in the United States by Vintage Books, a division of Random House,
Inc., New York, and simultaneously in Canada by Random House of Canada
Limited, Toronto. Originally published in hardcover in the United States by
Pantheon Books, a division of Random House, Inc., New York, in 2002.

Vintage and colophon are registered trademarks of Random House, Inc.

The Library of Congress has cataloged the Pantheon edition as follows:
Brooks, Rodney Allen.
Flesh and machines : how robots will change us / Rodney A. Brooks.
p. cm
ISBN 0-375-42079-7
1. Robotics.   2. Human-machine systems.   3. Artificial intelligence.   I. Title.
TJ211 .B697 2002
629.8'92—dc21
2001036636

Vintage ISBN: 978-0-375-72527-2

Author photograph courtesy of the author
Book design by Iris Weinstein

www.vintagebooks.com

Printed in the United States of America

11

Contents

Contents

Acknowledgments

I would like to thank all those who have helped me in getting these ideas, and getting them onto the page. My darling wife, Janet Sonenberg, encouraged me to do this and supported me throughout. This book would not be here if it were not for her. Also I must thank the following people:

- All my students over the years who have contributed to building robots and helping solve the problems of making life-like behavior from nonlifelike components.
- My sponsors from DARPA (the Defense Advanced Research Projects Agency) and the Office of Naval Research, and more

recently from NTT (Nippon Telegraph and Telephone Corporation), who have had patience and faith over the years that something good would come of my crazy ideas.

- Those at MIT who, while not quite believing in what I was doing, have at least let me do it, bravely standing by the idea of academic freedom.
- Marvin Minsky, who has inspired me since I was a teenager.
- The regulars at Dan Dennett's salon, including Marcel Kinsbourne, Marc Hauser, Ray Jackendoff, and Tecumseh Fitch, along with members past, fleeting, and future, who are always ready to discuss the nature of being.
- Annika Pfluger, my longtime assistant, who drew the figures for this book and, along with Sally Persing, tracked down obscure library references for me.
- The science fiction writers and moviemakers who inspire us, even when they are dead wrong, but especially when they challenge the human spirit to soar beyond itself.
- Colin Angle and Helen Greiner, who have stood with me for the past eleven years as we have tried to turn these dreams into reality out in the real world.
- My editors, Dan Frank and Stefan McGrath, who have provided many invaluable criticisms and suggestions.

Prologue

I have devoted my life to building intelligent robots, and these robots are just now starting to emerge from labs out into the real world. As these robots get smarter, some people worry about what will happen when they get really smart. Will they decide that we humans are useless and stupid and take over the world from us? I have recently come to realize that this will never happen. Because there won't be any us (people) for them (pure robots) to take over from.

Barring an asteroid-sized thwack that would knock humans back into pretechnological society, humankind has embarked on an irreversible journey of technological manipulation of our bodies. The first few decades of the new millennium will be a moral battleground as we question, reject, and accept these innovations. Different cultures will accept them at different rates (e.g., organ transplantation is currently routine in the United States but largely unacceptable in Japan), but our ultimate nature will lead to widespread adoption.

And just what are these technologies? Already there are thousands of people walking around with cochlea implants, enabling the formerly deaf to hear again—these implants include direct electronic-to-neural connections. Human trials have started with retina chips being inserted in blind people's eyes (for certain classes of blindness, such as macular degeneration), enabling simple perceptions. Recently, I was confronted with a researcher in our lab, a double leg amputee, stepping off the elevator that I was waiting for. From the knees up he was all human; from the knees down he was robot, and prototype robot at that—metal shafts, joints full of magnetorestrictive fluids, single-board computers, batteries, connectors, and wire harnesses flopping everywhere; not a hint of antiseptic packaging—everything was hanging out for all to see. Researchers are placing chips in animal, and sometimes human, flesh and letting neurons grow and connect to them. The direct neural interface between man and machine is starting to happen. At the same time, surgery is becoming more acceptable for all sorts of body modifications. I worry that I am missing the boat, carrying these heavy glasses around on my nose when everyone else is going down to the mall and having direct laser surgery on their eyes to correct their vision. And while all this is happening, cellular-level manipulation of our bodies is becoming real through genetic therapies.

We ban Olympic athletes who have used steroids. Fairly soon we may have to start banning kids with neural Internet connection implants from having them switched on while taking the SATs. Not long after that it may be virtually mandatory to have one in order to have a chance taking the new ISATs (Internet SATs).

With all these trends we will become a merger between flesh and machines. We will have the best that machineness has to offer, but we will also have our bioheritage to augment whatever level of machine technology we have so far developed. So we (the robot-people) will be a step ahead of them (the pure robots). We won't have to worry about them taking over.

Cambridge, Massachusetts

February 21, 2001

Flesh and Machines

1. Dances with Machines

What separates people from animals is syntax and technology. Many species of animals have a host of alert calls. For vervet monkeys one call means there is a bird of prey in the sky. Another means there is a snake on the ground. All members of the species agree on the mapping between particular sounds and these primitive meanings. But no vervet monkey can ever express to another "Hey, remember that snake we saw three days ago? There's one down here that looks just like it." That requires syntax. Vervet monkeys do not have it.

Some chimpanzees and gorillas have learned tens of nouns, a few ad-

jectives and a few verbs, expressed as signs or symbols. They have sometimes put these symbols together in new ways, like "water bird" to refer
to a duck. But they have never been able to say anything as sophisticated
as "Please give me the yellow fruit that is in the bag." That requires syntax. Chimpanzees and gorillas do not have it.

Irene Pepperberg has raised and trained Alex, the famous African
grey parrot for over twenty years, first at the University of Arizona and
more recently at the Massachusetts Institute of Technology (MIT). Her
results have been stunning and have rebuffed those who deny that any
animals other than humans have any component of language. Alex has
forced the naysayers to be much more careful in exactly what capabilities they deny for animals, even birdbrained animals. Alex is able to hear
words and speak them. He can answer questions like "How many round
green things are on the plate?"—even the first time anyone has ever said
the two adjectives "round" and "green" in sequence to him. Alex has
been heard to say as his trainer is evidently leaving for lunch, "I'm just
going out for lunch and will be back in ten minutes." But this is just Alex
parroting (and thus the word) what he has heard the trainer say in these
circumstances before. Alex has never said, "I see that you are going out
for lunch. I expect you'll be back in ten minutes." That requires syntax.
Alex does not have it.

Beavers gnaw down trees to build dams that back up rivers to produce pools where more of the grasses they eat can grow. Birds build nests
from twigs, grass stems, and other things they find, such as pieces of
wool. But neither of these are technologies in any true sense. If the situations are not completely stereotyped, then the animals are unable to generalize their evolutionarily built-in plan to accommodate the novel
circumstances.

Chimpanzees, on the other hand, use different tool sets in different
social groupings. The chimpanzees at the Gombe Preserve in Tanzania
select long twigs, strip off leaves and protrusions, and then dip them into
ants' nests and pull out edible ants. The chimpanzees in the Tai Forest of
the Ivory Coast do not know how to do this, but unlike those from Tanzania, this social group builds anvils from stumps and rocks, then uses a
wooden or rock hammer to break open nuts. The different groups learn
how to use these tools within their social groups, but there is very little
invention of new tool use. The same tool-use patterns have been ob-

served in the same groups for decades now with no innovations. And no chimpanzees have built tools that have survived over archeological times. Chimpanzees and other animals do not really have anything that remotely suggests human technology.

The only nonhuman tools that we have found preserved have been those of the Neanderthals. The Neanderthals were very close to humans in form, and there are current debates about whether humans and Neanderthals interbred. It is not known whether Neanderthals had language, or syntax, as there is debate over whether they had the vocal tract control necessary for speaking. Perhaps they had syntax in their sign language, as there is in modern sign languages of deaf people. They certainly had technologies to build complex hand tools, weapons, to manipulate fire, make jewelry, and prepare elaborate burial sites. Humans managed to kill off all of that species.

That leaves us humans. We alone on this Earth have syntax, and we alone on this Earth have technology.

For now.

What about our machines? Today there is a clear distinction in most people's minds between the robots of science fiction and the machines in their daily lives. We see 3CPO, R2D2, Commander Data, and HAL in *Star Wars, Star Trek,* and *2001: A Space Odyssey.* But these are not rivaled in any way by the capabilities of our lawn mowers, automobiles, or Windows 2000. There are the machines of science fiction fantasy, and then there are the machines we live with. Two completely different worlds. Our fantasy machines have syntax and technology. They also have emotions, desires, fears, loves, and pride. Our real machines do not. Or so it seems at the dawn of the third millennium. But how will it look a hundred years from now? My thesis is that in just twenty years the boundary between fantasy and reality will be rent asunder. Just five years from now that boundary will be breached in ways that are as unimaginable to most people today as daily use of the World Wide Web was ten years ago.

Throughout the recorded and archeological history of humankind, the way of life that the majority have followed has been dramatically shifted by technological revolutions. Technological revolutions are happening more and more often. The dawn of the most recent of these revolutions, the digital revolution, is within living memory of many of us alive today. The battles and restructuring of our world brought on by the digital revolution is raging around us at its full fury today. It is shifting wealth and power, and restructuring our cities and lifestyles.

We look forward to it running its course, and letting us catch our collective breaths. One wishes it were so simple. No, there is not another technological revolution brewing, about to inundate us with another tsunami that will toss our lives into disarray until we restructure them once again.

There are two. They are about to hit us almost simultaneously.

Technological Revolutions

Since before the start of recorded history technological innovations have transformed our lives as humans. In past times these transformations were gradual and happened over many generations. Now technological innovations can transform our lives many times within a single lifetime.

When I was a teenager, I received all scientific news by way of steamship on a three-month voyage. By 1984 I had cofounded a Silicon Valley company, Lucid, in Palo Alto and taken a position on the faculty at MIT. I was the chief compiler implementor for a Common Lisp product, and I telecommuted from my new home in Massachusetts. First, I used what I now call "Federal Express Net," by which we shipped 20-megabyte tape cartridges back and forth across the country daily by air. Later, I had an expensive leased land line that provided 19.2Kbaud access, and then a little before we arrived at Chapter 11 bankruptcy, I used the free Internet in the early 1990s to telecommute. Thus, I lived through three or four generations of information dispersion technology before I started to feel middle-aged, having experienced the history-in-microcosm of technological life transformations through a series of different machines. While these transformations happened at breakneck speed compared to those that engulfed our ancestors, in my case they were certainly rendered with much less violence or even discomfort. Just as our syntax defines us, so does our technology.

Early in mankind's history, we developed very simple machines, so simple that some people would be hard put to describe them as machines.

A club lets the wielder amplify the velocity of his hand to trade impact force for a reduced duration over what direct pressing of the hand would give. A hand plow concentrates force supplied over a large area to a small area at the point. Pulleys and screws let large and very large forces be applied over short distances by facilitating the hand application of a smaller force over a much longer distance.

All these machines provided physical couplings for transforming forces. All required direct human contribution of force, and all required continuous control by a human to operate. Every second, a human had to be in the loop providing moment-by-moment control to the machine.

By today's standards these early machines were not very complex. But they enabled the first great revolution in the way humans lived their lives. These simple machines enabled the *agricultural* revolution that began about ten thousand years ago (or 10,000 B.P.—before the present). This, the first of the great revolutions in human life was not at all abrupt; it took thousands of years to spread and become ingrained across the majority of the world's people.

During that transition other critical machines were invented, such as metal smelters and the wheel. The wheel first appeared around 8500 B.P., but it took another three thousand years before animal-drawn vehicles were invented by the Sumerians. With animals pulling them, our machines started to gain some autonomy. No longer did a person have to be in charge moment by moment. Rather, he could give a horse, say, an indication of which way to go, and the horse would handle the low-level details of traversing the ground and would avoid obstacles along the way.

The animal-drawn machines—carts and plows—brought about the second great revolution in human living in about 5500 B.P. The Sumerians were able to drain marshlands, irrigate new land, plow large areas for permanent cultivation, and transport agricultural products. These semiautonomous machines very slightly reduced the total manpower needed to produce food, and let just a few people take on specialized roles, such as priests, charlatans, artists, merchants, and most importantly, rulers. Thus was born the *civilization* revolution.

The next few thousand years brought incremental changes in technology, including water clocks, windmills, and other simple autonomous machines. These enabled towns and cities to become large, but still the majority of the population lived on farms.

It was not until the invention of the steam engine by Thomas Newcomen in 1705, followed by the replacement of charcoal made from wood with coke made from coal in iron smelters by Abraham Darby in 1709, that the stage was set for the *industrial* revolution. When James Watt in 1765 added a separate condenser cylinder to Newcomen's later 1712 design, the efficiency of steam engines went way up, and their size ultimately went way down. They were now strong enough to pull themselves around on wheels, and to economically provide energy to mills of all sorts.

The next two hundred years saw a complete remaking of patterns of human settlement. The majority of people moved to towns and cities. They performed two sorts of jobs. Some provided the minute-to-minute intelligence needed to run vast hordes of machines that processed materials into goods. The power systems were autonomous, but the control of the individual machines was not. People filled in the gaps, in autonomy, often in a very mindless, dulling sort of way, repetitively making simple decisions again and again, feeding machines as necessary. The others provided physical labor in places where machines could not do the job. People still loaded and unloaded material from trains, people laid bricks and nailed wooden beams to build houses, people paid attention to individual plants in nongrain agriculture, people washed and cooked, people sorted and redirected goods at all steps in the supply chain, people did all the accounting, banking, and taxing, and people directly controlled the machines of war. In short, people were still required to provide intelligence in environments that were not completely structured for machines.

The industrial revolution continued into the twentieth century, and as fossil-fuel-based power sources got smaller and more efficient, we moved from trains and buses to the individual automobile, which transformed city life first in the Western world and now throughout the world. Our central cities have become places that people go to during the day, and the suburbs have become the places where they go to sleep at night, constrained to be within a manageable commute from their places of work.

The revolutions—agricultural, civilization, and industrial—that we have talked about here did not happen overnight. They were spread over thousands of years initially, down to hundreds of years for the industrial revolution. We are at the tail end of that revolution, but already a visible new revolution has come along. A very different sort of technology is once again transforming our lives. The *information* revolution, powered by the microcircuit, is changing everything about the way we live. From within, it seems that it is perhaps just a quantitative change, not a qualitative change, but such is the nature of all profound changes that are not also disasters—they happen gradually and one must step outside to see the full dimensions of the changes.

The information revolution has changed our access to knowledge, or content, as it is sometimes called, in ways which may turn out to be more profound than Gutenberg's invention of the movable-type printing press in 1454. That technology enabled distribution of information to the masses and made it worthwhile to learn to read—now there was a constant supply of reading material beyond simple signs to identify a pub or a trading post.

The information revolution started with the invention of the telegraph in 1834. For the first time, information was available instantaneously at a remote location. The telephone democratized this capability, letting untrained people communicate throughout a city instantaneously. With the advent of electronic switches in the second half of the twentieth century, the telephone system was extended throughout the world, and without human help a person was able to dial almost any telephone in the world.

The information revolution is now in full swing, and it is once again transforming our cities. As with all inventions, the World Wide Web is not without its predecessors. In 1945, Vannevar Bush at MIT wrote an article, "As We May Think," in which he discussed possible future technological innovations, many of which have come to pass. Perhaps his most prescient statement was his inkling about computers when he reflected on the reliability of telephone exchanges: "The world has arrived at an age of cheap complex devices of great reliability; and something is bound to come of it."

He also talked about a hypertext knowledge system, called the *memex*, that might be a desk-sized device. He saw this as the best way to update the methods for disseminating scientific information that had been around largely unchanged for hundreds of years. Starting in the 1960s, Ted Nelson has designed many hypertext systems under the project title Xanadu that would change the way information was stored, referenced, and paid for, but he was not able to get beyond his purity of purpose to a practical distributed implementation. On the other hand, Tim Berners-Lee had to deliver something useful to his employers, and so when he invented the World Wide Web, he straightaway built it and laid it on top of the nascent Internet in the early 1990s. In a lightning-fast spread of the technologies and protocols, the world suddenly had instantaneous availability of almost any information anywhere. We no longer need to travel to the site of information. We can demand, through a few clicks of a mouse button, that it come to us.

This is rapidly changing how much human mediation there needs to be in our access to information. One of the functions left for people in their move to cities during the industrial revolution is being depleted. There is no longer as much need for people in management of the supply chain, and in accounting, banking, and taxing. People are indeed buying books, groceries, appliances, furniture, automobiles, stocks, and travel over the Internet. There is a whole new economy that has formed around online auctions. People who could not afford to sell things before are participating in booming markets for collectible items. They are doing business directly all over the world, without regard to geographical location.

This is already having impact on our cities. As Bill Mitchell, dean of architecture at MIT, likes to point out, the information revolution is converting neighborhood bank branches in U.S. cities into Starbucks coffee stores. It is also having an impact on the way we work—more and more people are now working at home, telecommuting. As higher bandwidths become available, there will be less and less need to be collocated with the place of work, or purchase, and more and more people will move not just tens of miles from their company, but thousands. The constraint of a reasonable commute will be gone and our cities will evolve.

While we are just now heading into the middle of the information revolution, two new revolutions are coming up fast upon us.

The *robotics* revolution is in its nascent stage, set to burst over us in the early part of the twenty-first century. Mankind's centuries-long quest to build artificial creatures is bearing fruit. Machines are now becoming autonomous in the areas that bypassed them in the industrial revolution. Machines are starting to make the judgments and decisions that have kept people in the loop for the last two hundred years. There will soon be less need for people to engage in the moment-to-moment control of manufacturing machines, and we are starting to see intelligent robots that can operate in unstructured environments, doing jobs that are usually thought to still require people. But these robots are not just robots. They are artificial creatures. Our relationships with these machines will be different from our relationships with all previous machines. The coming robotics revolution will change the fundamental nature of our society.

Hot on the heels of the robotics revolution is the *biotechnology* revolution. We might think that biotechnology is already here. But we have seen only the merest shadows of its reality. It will transform the technology not just of our own bodies, but also that of our machines. Our machines will become much more like us, and we will become much more like our machines. The coming biotechnology revolution will change the fundamental nature of us.

Further Reading

Bush, V. 1945. "As We May Think." *Atlantic Monthly,* July. Available online at http://www.isg.sfu.ca/~duchier/misc/vbush/vbush.shtml

Griffin, D. R. 1992. *Animal Minds.* Chicago: University of Chicago Press.

Hauser, M. D. 1997. *The Evolution of Communication.* Cambridge, Mass.: MIT Press.

Nelson, T. 1974. *Computer Lib/Dream Machines.* Peterborough, N.H.: Bits Inc., Rev. ed. published in 1987 by Tempus Books of Microsoft Press, Redmond, Wash.

Pepperberg, I. M. 2000. *The Alex Studies: Cognitive and Communicative Abilities of Grey Parrots.* Cambridge, Mass.: Harvard University Press.

2. The Quest for an Artificial Creature

ankind has always used the technology of the day to make images of animals, real and imagined, and of people. The cave paintings of Chauvet Pont D'Arc in France, from 30,000 years before the present (B.P.), depict rhinoceroses, bears, lions, and mammoths, along with pictures of human hands and other parts of human anatomy. In the caves of Lascaux, also in France, from 17,000 years B.P., there are pictures of people hunting bison.

Besides pottery to hold food and liquids, around 7000 B.P., Europeans and Chinese started making and baking small clay animals and people. Earlier people had carved figurines from wood and stone. As soon as metal became smeltable, about 4000 years B.P., there appeared metal animals and metal models of people. Such static figurines have played artistic and religious roles in human affairs since before recorded history.

These paintings and sculptures were representations of animals and people, but they did not *act* in any way. Mechanical technologies were needed to give this aspect of realism to physical artificial creatures.

Mechanical Creatures

When hinges and wheels were invented, they soon found their way into articulated figures, and animals that could be pulled along the ground by a string. Both a pig and lion made in such a way from limestone were found at the temple of Susa in Persia, constructed 3100 years B.P. By the height of the ancient Egyptian civilization there were articulated (and sometimes articulate too, through the deceit of sound passages opening into the mouths) statues that were secretly controlled by priests. At Thebes an articulated statue of Ammon chose the new king by reaching out an arm, under the unseen control of a priest, as the male members of the royal family paraded in front of it.

In Greek and Roman times clepsydrae, or water clocks, were developed. The earliest water clocks, like that found in the tomb of Amenhotep I, buried around 3500 years B.P., simply accumulated dripping water. By 2,000 years ago the Greeks and Romans had added pistons, gears, and ratchets so that precursors to modern clock faces could be added. By the year 100 of our current calendar, Hero of Alexandria had built pneumatically operated human forms—he also wrote extensively about clepsydrae, and even built a primitive steam engine.

After that, through most of the Middle Ages, not much happened in terms of mechanical or indeed any technological advances. Despite claims that Roger Bacon built a mechanical talking head in the thirteenth century, the evidence is scant; more likely, along with his telescope and other inventions, it was a product of ingenious and accurate speculation rather than actual construction.

From the early-to-mid fourteenth century into the seventeenth century, there was a passion for building clockwork simulations of the heavens. They not only displayed the phases of the moon and the constellations of the zodiac, measured out in real time, but they also struck individual hours. In some it was animated figures, little humanoids, who did the striking. This technology eventually became affordable and was produced en masse after the first cuckoo clock was designed by Anton Ketterer around 1740.

In the meantime, the incomparable Leonardo da Vinci, following his detailed studies of human anatomy, designed a mechanical equivalent to a human—a humanoid robot—early in the sixteenth century. Like many of his greatest inventions, this one was not constructed in his lifetime, and indeed still awaits an instantiation.

When Louis XIV of France was a child, a craftsman named Camus built an elaborate set of toys for him. They consisted of a coach and horses with footmen, and occupants in the coach that were all articulated. Later in his reign, in 1688, a General de Gennes constructed a peacock that walked and ate. This perhaps inspired Jacques de Vaucanson, who in the eighteenth century built the most famous mechanical creatures.

Vaucanson is best known for his mechanical duck with its intricate wings, modeled very carefully after the real thing. It paddled and quacked and extended its neck to take food and water. Vaucanson's duck ate and also defecated, although there has long been debate over whether the digestion was automated or was through trickery.

Vauconson's other creations included three humanoids. One was a mandolin player. It sang and tapped its foot as it played. Another was a piano player that simulated breathing and moved its head, and a third was a flute player. The reports of these creations all stress how lifelike they were. But one must keep in mind how the standards of lifelike change over time. The very first black and white movie of an oncoming train had patrons running in fear from the Paris theater in which it was shown. Now we hardly flinch at the most realistic three-dimensional movie that can be created, and old black and white movies seem to us to be slightly abstracted representations rather than lifelike reality.

In any case, the works of Vaucanson inspired others, and there were soon a series of mechanical humanoids and other animals built and displayed throughout Europe. Later in the eighteenth century Pierre Jacquet-Droz and his son Henri-Louis built a number of humanoids, including a female organ player. This creature simulated breathing and gaze direction, looking at the audience, her hands, and the music. Henri Maillardet in 1815 built a body that could write script in both French and English and draw a variety of landscapes. It is now housed in the Franklin Institute in Philadelphia.

These automatons were impressive to the people of the time, and the more sophisticated of them remain awe-inspiring when seen in operation in museums today. They could act in the world. But these artificial creatures lacked spontaneity. They did exactly the same thing every time they were activated. They did not *respond* to their environment in any way. Electronic technologies were needed to give this extra aspect of realism to physical artificial creatures.

Electronic Creatures

During the 1930s and 1940s the first digital computers were built. They used as their switching elements either electromagnetic relays or vacuum tubes.

Electromagnetic relays consist of an iron core with a wire wrapped around it thousands of times, like cotton on a bobbin. When a current is passed through the wire, a magnetic field is induced and this pulls a switch mounted at the end of the core. Relays were an abundant commodity item because of their role in the central offices of telephone companies, where they were the mainstay of the equipment that automatically routed dialed telephone calls. Such equipment had supplanted the human operators who physically rerouted calls in the early days by plugging cords between the incoming line and the destination line.

Vacuum tubes had evolved from the technology developed to make incandescent lightbulbs. A glass tube contains a glowing filament, tuned in this case to give off electrons rather than photons, a plate to collect them and form a circuit, and a small metal grid in between. When a voltage is applied to the grid, the flow of electrons between the filament and the plate is modulated and so can act as an electronic switch. Vacuum tubes were abundant, as they were the main element of radios and later televisions.

Various individuals and teams in Germany, Great Britain, and the United States started to put these elements together to form primitive digital computers. Development was spurred during the Second World War as computers were used to crack intercepted coded enemy messages. But the machines were enormous in size. In 1950, ENIAC at the Moore School of Electrical Engineering at the University of Pennsylvania was the first "modern" electronic computer with the essential features we see in today's machines. But it occupied three rooms, had 18,000 vacuum tubes, and had a mean time to failure of only twenty minutes. Vacuum tubes always eventually fail, just as their predecessors, the incandescent lightbulbs, still always fail today in our homes. The filament eventually wears out. Making vacuum tubes with longer lifetimes was a major area of development for their manufacturers. But they still have lifetimes of only a few thousand hours at most. A radio with four or five vacuum tubes could therefore play for hundreds of hours before one of its tubes burnt out and a trip to the radio shop was necessary to buy a replacement. Most people were able to make this sort of repair themselves. But for a digital computer with a total number of vacuum tubes bigger than the number of hours of average lifetime of each one, there was sure to be a failure of one of the tubes after less than an hour's operation. And then there was the task of tracking down which of the thousands of tubes had failed. These early computers, though somewhat less powerful than one we find in a $5 watch these days, were enormously expensive and were kept busy on military problems. They were not the domain of tinkerers intent on building artificial creatures, and even if people were thinking about it, certainly no one had really figured out how to program them to be the nervous systems of an artificial creature.

The individual switching elements were available at moderate prices, however. A switching element was what had been missing in all previous attempts to build artificial creatures. The switching elements gave artificial creatures the ability to respond to their environment. Both the electromagnetic relay and vacuum tube, however, provided a means for changing behaviors based on signals received from sensors. Given these new technologies, there were quite a few attempts at building artificial creatures. Most of them are not documented and many were not at academic institutions. The noted philosopher Daniel Dennett came across a marvelous creation, a robotic dog in a Paris secondhand store. He has spent many unsuccessful hours trying to track down its history, and even more very enjoyable hours figuring out how it works and restoring parts of it to working order. The dog can bark and move on its wheels, and seems to be sensitive to light.

A few people, such as Thomas Ross, R. A. Wallace, and the famous Claude Shannon, had built robots that found their ways out of mazes, but these robots were not very animal-like in the way they operated. But one attempt, which was quite successful at producing animal-like behavior and was well documented, was by William Grey Walter.

W. Grey Walter, as he was known, was a somewhat eccentric American who was head of the Physiological Department at the Burden Neurological Institute in Bristol, in the west of England. Most of his 170-some papers were devoted to the science of electroencephalography. An electroencephalograph noninvasively measures human brain waves by picking up electrical activity with electrodes taped to a patient's scalp. This is very much an observational science—it is not one where you can go in and change the insides of a person's head and see what happens. Walter realized that if he could build electromechanical models of animals he would be able to actively experiment with their innards and perhaps get new insights into how nervous systems worked.

Sometime around 1943, Walter, with the help of his wife Vivian, started building small robots. They used both relays and vacuum tubes as the switching elements, gears from old gas meters, and simple electric motors as actuators. Walter was interested in building artificial creatures that not only displayed *spontaneity* but also *autonomy* and *self-regulation*.[1] By 1948 they had successfully constructed the first of a series of machines that he dubbed with the mock-biological name *Machina speculatrix*, and they were displaying them to the *Daily Express* newspaper by December 1949. In 1950 and 1951, Walter published two short articles on them in *Scientific American*. The first of these two papers was titled

1. See chapter 5, "Totems, Toys, and Tools," in Walter's 1953 book *The Living Brain*.

"An Imitation of Life," and from his other writings it is clear that Walter was taking delight in the way his robots appeared to respond to the world like animals. Like Vaucanson and others before him, Walter had become romanced by the idea of making a mechanical version of an animal.

Walter's robots were about the size of a shoebox and looked a little like giant snails—a rounded plastic shell with a little head sticking up at the front. They are often referred to as both turtles and tortoises, although Walter himself called them tortoises. In the manuscript of a talk, Walter quotes Lewis Carroll from *Alice's Adventures in Wonderland*. "We called him Tortoise because he taught us," said the Mock Turtle angrily: "really you are very dull!"

The tortoises each had two electric motors, one for steering and one for motive power, as actuators, and the mechanical systems were built out of gears from old clocks and gas meters. There were three wheels in a tricycle arrangement, with the front wheel doing all the work—it was the drive wheel and the steering wheel. One motor drove it so that the tortoise could move, and another motor turned it continuously without any steering limits, so that it could face backward or any other direction. Under normal conditions the steering motor would drive continuously at moderate speed so that the steering column was scanning continuously.

The electronics were based on miniature vacuum tubes and electromagnetic relays. All the robots had a single bump sensor, implemented as a hanging skirt in such a way that a bump from any direction closed a switch. They also all had a light sensor mounted on the steering column, so that the tortoises could determine the light intensity in their then current direction of travel, as the steering motor scanned about.

Some of the robots had additional sensors, such as microphones. Additionally there was a special hutch, with a light inside and visible from its entrance, and a special recharging mechanism so that if a tortoise entered it would be coupled to the recharging station.

The first class of tortoises built, named *Machina speculatrix* because they seemed to exhibit "exploratory, speculative behavior that is so characteristic of most animals," had only two vacuum tubes, two sensors (light and bump), and the standard two actuators, or motors.

The simplest behavior exhibited by the tortoises was their attraction to light. Their steering motor would stop scanning whenever there was a moderate light right in front of its sensor, so it started heading toward the light source. Perhaps it would accidentally veer away as the rear wheels skewed its direction, but then it would scan again, soon pick up the light again, and gradually make progress toward the light.

The tortoises also had another fundamental behavior. When the bump sensor was activated, then for a time the light sensitive effects were inhibited and full power oscillations were provided to both motors, causing the tortoise to move in "a succession of butts, withdrawal and sidesteps until the interference [was] either pushed aside or circumvented."

The circuits were arranged to be very nonlinear. Under moderate levels of light, the steering motor stopped scanning and the tortoise moved toward the light source. When the standard light source used in the experiments was only about 15 centimeters from the tortoise, a relay kicked in and the scanning-steering motor was run at double speed, making the tortoise abruptly turn away. Normally this reaction was initiated before the tortoise found its way into its home hutch with the light inside. As the batteries ran down on the robot, however, the sensitivity to light decreased and the robot would find itself all the way into the hutch before the flee mechanism fired. As soon as it was inside and connected to a power source, the motors were switched off until the batteries were fully charged. At that point it again became very sensitive to light and found its way out. The fact that the tortoise "decided" to leave the hutch was an instance of an emergent behavior that nonlinearly coupled other more simple primitive behaviors in a particular environment in such a way that another behavior, naturally described in different terms, came about.

There were other cases of emergent behavior too. The tortoises were equipped with a lightbulb that indicated when the steering motor was switched on. When the tortoise came across a mirror, there was a coupling set up through the external world and an oscillation started up. The robot would be attracted to its own reflected light, which immediately switched off the steering motor and the light, whence it was no longer attracted and so the light came back on, and so on. Other interesting things happened when two tortoises met each other head-on.

Later Walter built a more elaborate circuit for his tortoises. There is some debate over whether the circuit was actually contained inside the tortoises as his writing implies, or whether it was an external device connected by wires. The circuit would enable the tortoise to learn things. He called the new tortoises *Machina docilis*.

Walter wanted the tortoises to be able to learn conditioned reflexes. The archetypical experiments which demonstrated that animals could learn in this way were carried out on dogs by a Russian physiologist named Pavlov, early in the twentieth century. When food enters the mouth of a dog, there is a flow of saliva. This is a simple reflex, apparently hard-wired into the animal. The food is referred to as an unconditioned, or specific, stimulus. Over a number of trials a bell was rung whenever a dog was given food. The sound of the bell is referred to as the conditioned, or neutral, stimulus. Before long, the dog learned the association of the two stimuli and started to salivate when the bell was rung, even when there was no food.

Machina docilis exhibited conditioned reflex learning. The unconditioned reflex was evoked by kicking the tortoise. This triggered its bump sensor, and its steering motor ran at double speed as it moved away rapidly. Walter used a pure 3,000Hz tone as the neutral stimulus—it was much easier to detect than to build a sensor and perception system that could recognize all manner of bells being rung. Before long, the tortoises would learn to flee on the sound of the tone. And the tortoises could also learn associations with other unconditioned stimuli, such as shining light at their light sensors.

Walter noted that the behavior of the tortoises was "remarkably unpredictable." There were many sources of subtle variations. For example, there were changes in perceived light level due to small changes in voltage for the actual light source, and due to even smaller changes in voltages in the detection circuits as the motors drew more and less current in response to varying forces applied to the tortoise as the steering angle changed. These microeffects combined in such complex ways that it was very hard to predict the behavior of the tortoise.

There are a number of lessons to be learned from these early experiments with artificial creatures. The most important is that even a seemingly very simple creature can have extremely complex behavior in the physical world because of the way that small variations in what is sensed, and how the actuators interact with the world, can change the actual behavior of the system. Second, even simple creatures, with just a few nonlinear elements, can end up producing a wide variety of emergent behaviors, where multiple elementary behaviors couple with the environment to produce behaviors that are more than simple strings or suppositions of the elementary behaviors. An observer finds it easier to describe the behavior of the tortoises in terms usually associated with free will—"it decided to go into its hutch"—rather than with detailed mechanistic explanations of the particular unknowable details of exactly what its sensors reported when. Third, detailed accounts of Walter's experiments make clear how difficult it is to implement a theoretical psychological or neurological construct on a physically embodied system. The details are often not important at the level of distinction being made in the theories, but they are essential in order to actually get instances of the theory coupled into all the other constraints that the body and the world provide. Even though Walter was using a much simpler audio stimulus for his tortoise than Pavlov had used for his dogs, a pure tone rather than a bell ringing, Walter had to make seven separate extensions to Pavlov's description of the process in order to make it work. For instance, he had to explicitly incorporate a circuit which held in memory that the tone had recently been heard, so that information was still available when the shell was kicked. Walter's experiments required very careful design in order to emulate the learning capabilities of simple animals.

Digital Technologies

Transistors, solid-state switching elements made from germanium or silicon, let people build reliable computers that could run for hours and even days without a hardware failure. From the late 1950s these computers started to revolutionize business accounting and scientific investigations. By the late 1960s they were built around integrated circuits, the first chips. These days the central processing unit is a single chip, or microprocessor, but in the sixties the CPU was still a large collection of chips occupying one or more racks of electronics. A computer that was used for serious research occupied a large room, was operated by a full-time staff of a few people, and at a really big institution had perhaps a megabyte of main memory. Memory at that time was primarily made up of magnetic *cores*. Each individual bit (there are 8 bits in a byte of memory) was represented by an individual ring of metal, about a millimeter in diameter. These cores were magnetized by passing currents through multiple wires threaded through them, and the direction of the north pole of this tiny magnet represented either a zero or a one. Another *sense* wire passed through the core, and the directionality of the magnetic field affected a test current flowing through the wire. Adding in some redundancy to correct the high error rate of magnetic cores, a megabyte of memory required about 10 million of these little donuts, each with three or four wires threaded through. Apart from the physical volume of the memory, the power requirements for all the circuits, and cooling them, was enormous. These computers were too big to put on board a robot, but people started thinking about mobile robots with their brains off board—a radio and TV link from a brainless set of wheels, motors and sensors, to a large mainframe computer that did the thinking and sent commands back to its physical avatar, the shell of a robot, in the physical world.

The most celebrated such robot was Shakey, built at the Stanford Research Institute in Menlo Park, California. (As a result of its divestiture from Stanford University during Vietnam, the largely Defense Department–funded Stanford Research Institute is now known as SRI International.) Nils Nilsson led a team of researchers from 1968 to 1972 who built and programmed Shakey, and made many foundational contributions to artificial intelligence.[2] Nils is a tall white-haired man whose physical appearance and quiet and good humored contemplative demeanor confirm his Scandinavian heritage. Shakey's demeanor mirrored that of its creator. Shakey, about the size of a small adult, lived in a set of carefully constructed rooms, and sensed large colored blocks and

2. Surprisingly there is no generally accessible account of the Shakey project by its creators. There were a number of scattered research papers at conferences, and Shakey is mentioned in many textbooks as the canonical reference for the origin of many subfields of classical artificial intelligence research. It is also referred to in many popular books. The only technical integrative account is a collection of papers and recollections put together by the project's leader as an SRI technical note.

wedges. It could push them along the floor. Typically, Shakey, so named for the way its camera and TV transmitter mast shook as it moved, would be commanded to go to a particular room and push a particular colored block to another room. Along the way it might encounter a doorway blocked by a brightly colored wedge that it needed to move out of its way to proceed with its plan. Eventually Shakey would get to its goal a few meters from where it started and carry out its task—six or eight hours after it started. Most of the time Shakey, the robot shell, sat idle while its remote brain contemplated a long series of moves to accomplish its ultimate goal. Its on-board camera transmitted a standard TV signal that was received and digitized every few minutes by the main computer. Its bump sensors and wheel velocity sensors sent reports a few times a second via a radio system. And the main computer would very occasionally send commands to turn its wheels or to pan or tilt its camera.

Shakey was hardly an artificial creature in the way that Grey Walter's creations were. It used a completely different approach to being in the world. The tortoises had no *a priori* knowledge of the particular circumstances they would find themselves in when they were switched on. They had to sense everything themselves. For Shakey, however, the researchers made a complete two-dimensional map of the world before Shakey was started up, and stored it in the computer's memory. Implicit in Shakey's programming was the assumption that the floor would be perfectly flat, and that nothing could sit on top of anything else, so two dimensions were sufficient. As Shakey moved about, it modified this model, changing its internal belief about where the colored blocks were on the basis of its visual perceptions. Shakey, however, hardly acted like an animal. It sat for minutes at a time while its remote brain computed. It clearly had no sense of the here and now of the world—if someone came and moved things around while it was "thinking," it would eventually start up again, acting as though the world was still in the same state it had been before the perturbation. Shakey used reasoning in situations where real animals have direct links from perception to action. It was designed on the premise that its perceptual computations could maintain an accurate model of the world, but that was so technically difficult that its designers were forced into the deceit of making the world very simple so that this was possible.

By the late 1970s there were three well-known mobile robot projects that, one way or another, followed from the Shakey work.[3]

At the Jet Propulsion Laboratory (JPL), part of NASA, in Pasadena, California, there was a robot that was used for experiments aimed at building a rover to operate on the surface of Mars. This was a large flatbed robot about the size of a small car, with a manipulator arm mounted on it. Very little work was done on navigation, and instead most of the published papers concentrated on the computer-vision-guided picking up of rocks.

At the Laboratoire d'Analyse et d'Architecture des Systèmes (LAAS) in Toulouse, France, there was a three-wheeled mobile robot named Hilare. The two mainstays of the Hilare project were Georges Giralt, an affable Spaniard, and Raja Chatila, an intensely intellectual Syrian, who had both transformed themselves to be quintessentially French. Hilare was a little bigger than a shopping cart, and used sonar and a laser range finder to build two-dimensional models of its world—a world of movable plywood walls that its creators would set up as a challenge for it to navigate through. And this is exactly what Hilare did. It would scan its sensors accumulating evidence of where the walls were, plan a collision-free path in two dimensions based on its best current beliefs, and move a little. Then it would integrate the evidence newly before its sensors with its previous view of the world to get the most consistent model that it could. Then it would plan again and move a little farther toward its goal. Like Shakey, Hilare did not qualify as an artificial creature.

During the rush to get to the Moon in the Apollo program, NASA sent unmanned landers called Surveyors. One of the most important pieces of information gleaned from these landers was that the Moon's surface was solid enough to land on without being swallowed up. During the crash development of the Surveyors there was for a little while a plan to send a mobile robot to test the traversability of the surface—there was some fear that the astronauts might safely land on the Moon but then not be able to move about. NASA got the Mechanical Engineering Department at Stanford University to build a four-wheeled vehicle, to test how well it could be driven from Earth via radio signals. Other re-

---

3. I would not be surprised if there was also contemporary significant similar work somewhere in the Soviet Union, but I have not been able to uncover it.

search teams were working on making a robust rover, which eventually turned into the electric cars driven by astronauts on *Apollo 15, 16,* and *17.* At Stanford the issue was the controllability of a car from the Earth with the lag time that radio waves would have of two and half seconds round trip. So a rather simple device, called the Cart, was built. It had four bicycle wheels and a drive system that used a bicycle chain. The front two wheels were steered much like an automobile. The Cart was roughly the size of a card table and looked a lot like one with bicycle wheels on its legs. The plan to send the robot car to the Moon was soon dropped, and so the Cart languished at Stanford. Before too long it was adopted by some hackers at the Stanford Artificial Intelligence Laboratory, and it kept three generations of graduate students busy during most the 1970s.

The Stanford Artificial Intelligence Laboratory (SAIL) had been founded by visionary John McCarthy in 1963. John had earlier cofounded and codirected the MIT Artificial Intelligence Laboratory (which I now direct) along with Marvin Minsky. John had high ambitions for artificial intelligence, a name he had coined for a field that he had started by running a six-week workshop at Dartmouth College during the summer of 1956, "The Dartmouth Summer Research Project on Artificial Intelligence." He set up SAIL in the D. C. Power Laboratory Building,[4] a wooden semicircular building in the hills a few miles above the Stanford campus, near the San Andreas fault.

John believed in reason, and brought that and very little else to his own human interactions. He was well respected, and well liked, almost worshiped, but remote. His alter ego was Lester Earnest, who ran the day-to-day operations of SAIL. Lester was a tall, affable Ph.D. who had played varsity football for Cal Tech. Lester doled out both office space and research money. Under McCarthy's guidance, Lester oversaw the gradual construction of the world's most unique and advanced mainframe computer, all the time under continuous use by tens of graduate students as it was being built. At the time that most university computers still used punched cards or at best character-only video terminals, the SAIL computer had full bit-mapped graphics, video switchable to

---

4. Named for Donald C. Power; the joke was always where to find the A. C. Power laboratory.

any terminal, digital sound at every terminal, and a laser printer that could produce black and white graphics. Lester had encouraged a sense of play in the students at the lab, and they had responded by producing Space Wars, the first video game; the first electric robot arms; some of the first computer-generated music; the first full-screen editors; and some of the first typesetting systems that used variable-width fonts. SAIL was a hotbed of innovation and intellectual ferment. But SAIL was certainly not beyond making mistakes. One day a young fellow named Steve Jobs came by to see John McCarthy with a computer built on a wooden base, touting it as the future. He was turned away. John made up for it a couple of years later by sponsoring the Stanford University Network project that turned into Sun Microsystems.

It was in this environment that three graduate students, one after the other, worked under John McCarthy's supervision trying to make the Cart do something intelligent. McCarthy was not impressed by the levels of emotion and illogic that were displayed by humans driving cars. He wanted to replace human drivers by safe computer drivers.

The first student was Rodney Schmidt. In 1967 he and McCarthy started to think about how to make an automatically driven automobile that would be able to operate in traffic on unmodified roads and to get all the necessary information for driving from its own sensors. When Schmidt turned in his thesis four years later, he reported on how the Cart was able to follow a white painted line along the ground for about twenty feet. The problems had been much tougher than expected, but Schmidt's real innovation was that the Cart was situated in its environment. It did not have a full internal world model, rather it sensed what was right in front of its eye and responded to that—following the white line as it twisted and turned. And it operated outdoors rather than in an antiseptic indoor environment. It was the first inkling of how a robot might ultimately be like an artificial animal, responding to its environment in real time rather than contemplating its every move—sensory-motor loops rather than cognition.

The lure of reason and logic was strong, however. The next student to take over the Cart was Bruce Baumgart. Perhaps stimulated by the hilly three-dimensional environment around SAIL, Bruce decided that it would be necessary to have three-dimensional models of the world. This was before three-dimensional graphics systems were commonplace, so Bruce had to invent many of the techniques that are now used to produce much of the imagery we see in movies, on television, in advertising, and in video games. This side problem turned into a major enterprise, and while extremely successful and influential, it meant that Bruce never really got the Cart to do much of anything.

The Cart was inherited by Hans Moravec. Hans was a true eccentric. Brilliant, innovative, and nuts. He was a tremendous influence on my life, once I got to Stanford and met him.

I had grown up in Adelaide, South Australia, a quiet city nestled between wine-growing regions to the north and south, and situated between Perth (2,000 miles to the west) and Melbourne (500 miles to the east). It was an isolated place at the end of the technological Earth, geographically as far away from the innovations of Europe and North America as one could get. We got books and magazines from England, our cultural homeland, three months after their publication date— that was how long it took ships to get from there to us. I grew up a nerd in a place that did not know what a nerd was. I would stare through the window at the IBM mainframe in the financial district of the city, lusting after the technology.

I spent my childhood building things, and by the time I was ten years old I was starting to figure out electricity enough to try to build my own computers, on a budget of around ten cents per week. A little later my father was able to bring home scrapped modules from the defense laboratory where he worked as an electrical technician. I built machines that could play simple games and, ultimately when I was twelve, one that could play tic-tac-toe flawlessly. Unfortunately, due to the limited number of components I had, the machine was restricted to playing in only one of the eight possible rotations and reflections of the board, and none of the adults around me would ever accept that I wasn't cheating them somehow with this restriction. I came across a Pelican edition of Grey Walter's book, and tried to build my own version of *Machina speculatrix,* using transistor technology rather than vacuum tubes (I had gotten plenty of nasty high-voltage shocks already building a vacuum tube oscilloscope of my own design). The subtleties of the original electronics were a little beyond me, but I did manage to get my first robot, Norman, to the point where it could wander around the floor, respond to light, and bumble its way around obstacles.

I then spent six years at the Flinders University of South Australia, studying classical mathematics with refugees from the Prague Spring of 1968. But all along I managed to get unrestricted access for twelve hours every Sunday to the university's mainframe computer—a 16-kilobyte machine. This was my heaven. I developed a computer language especially for artificial intelligence, and made an interactive interface for it on the main control console. There I wrote programs that proved theorems, did mathematical symbolic integration problems, understood some aspects of English, and learned how to play games. Eventually, I realized that this is what I should do with my life. I quit a mathematics Ph.D. halfway through and managed to secure a research assistantship, by mail, at the Stanford Artificial Intelligence Laboratory. They took only three new students in 1977, the year that I joined, and I still marvel at the incredible luck that let me get to Silicon Valley just at the time everything started to really blossom there. A kid from nowhere who knew nobody. What was the admissions committee thinking that year?

At SAIL, I soon met Hans Moravec. He had been there a few years already and was the master of the more arcane aspects of the SAIL computer. He had spent the last six months secretly living up under the roof of the lab in a bedroom he had constructed in the rafters. He had not left the lab during that whole time—his friends went grocery shopping for him, and he bathed in a shower that was down in the basement. This was all pretty wild stuff for me and my colonial middle-class background. On top of that Hans was full of fantastic ideas, ranging from skyhooks to get into orbit cheaply, to massively parallel computers, to treelike robots, to downloading human minds into silicon. His work with the Cart, however, was a little more pedestrian.

Hans was convinced that as the first step toward making the Cart act intelligently in the world, he needed it to make an accurate three-dimensional model of the world.[5] His first attempt at Stanford used computer vision to sense the world in three dimensions. He had to invent a whole lot of new techniques. His algorithms found visually distinctive points out in the world, and by matching how they appeared

---

5. He has carried this conviction for twenty-five years, and his latest research project at Carnegie Mellon University is his umpteenth implementation of an even better three-dimensional reconstruction program. I have tried to convince him over the years that

from two different cameras, or eyes, he was able to compute their direction and distance, much like we humans use our stereoscopic vision. Strictly speaking, Hans's version of the Cart did not use two cameras. Rather it used a single camera that he slid from side to side, to nine positions, and compared the images from each position. It was thus nine-eyed stereo vision, where a single camera was time-multiplexed to play the role of nine eyes. The Cart would enter all these points in a model, then compute the optimal path to its goal while avoiding all the obstacles. Then it would blindly move forward about one meter and look at the world again. It would integrate the new obstacle points in with its old world model, trying to get the most consistent match between successive observations.

The Cart transmitted camera images to SAIL's main computer using a TV transmitter (it was licensed as Bay Area TV Channel 22), and received back radio commands to steer and drive. The Cart made its forward lurches about once every fifteen minutes. Once every fifteen minutes if SAIL's main computer wasn't busy running jobs for other people. During the peak computer use times in the midafternoon, this fifteen minutes could easily stretch to three hours or more. There were a couple of other problems. First, the Cart's batteries ran down after about six hours of operation, and it was too flimsy to handle any more battery weight. Second, in order to make the algorithms run as quickly as they did, he had built in assumptions that the world remained static for the duration of a run. But SAIL was a hive of activity, and people, furniture, and doors, etc., were always moving about. Except at night. After midnight most students left, and there was hardly anyone around until about six in the morning, when the computer music students from CCRMA (the Center for Computer Research in Music and Acoustics) showed up to work on their compositions on the computer. During the summer of 1979, Hans ran the Cart just about every night through a large open room in the interior of SAIL. I was personally experimenting

(1) animals, including humans, do not make accurate three-dimensional maps and are still able to act intelligently in the world, and that (2) once he has these maps he is going to have to do something clever with them, so perhaps it is worth thinking about that problem a little now in order to guide requirements on the characteristics of the maps. I have yet to succeed.

with a twenty-eight-hour day—twenty hours of work, followed by eight hours at home—that summer, so I helped out a lot of those nights. We would set up an obstacle course of chairs, and cardboard polyhedra and trees that we had made, give the Cart a goal just beyond the far wall about 60 feet away, and then let it run. During its fifteen-minute periods of blindness while it digested its last observations, we could move through the room and make any adjustments that were necessary.

The Cart was generally able to find its way from one end of the room to the other over a six-hour period, but it often got confused by the cardboard trees and polyhedra. They did not have many visually distinctive features, so the Cart was often blind to them. We dressed them up with leis from a Christmas party and they became much more visible and avoidable. The culmination of these experiments came on a Saturday in late October when Hans was given sole use of the mainframe for the day, and he set up for a Cart run outdoors. It was a great spectator event, although even slower than a cricket match, as the Cart tried to navigate between the same chairs and cardboard models now placed out by the loading dock of the D. C. Power Lab. Unfortunately one of Hans's key assumptions was violated by the outside world. The highly distinctive shadows cast by all the objects on the ground were not static. In fact, they moved quite a bit in the fifteen-minute intervals between observations of the world. The Cart got very confused about its model of the world, and Hans had to run out and fix the world occasionally so that the internal model was a little more accurate—it was much easier to change the real world than to change the Cart's internal model of that world.

Despite the serious intent of the project, I could not but help feeling disappointed. Grey Walter had been able to get his tortoises to operate autonomously for hours on end, moving about and interacting with a dynamically changing world and with each other. His robots were constructed from parts costing a few tens of dollars. Here at the center of high technology, a robot relying on millions of dollars of equipment did not appear to operate nearly as well. Internally it was doing much more than Grey Walter's tortoises had ever done—it was building accurate three-dimensional models of the world and formulating detailed plans within those models. But to an external observer all that internal cogitation was hardly worth it. It brought to mind the old saw about whether a tree falling in an uninhabited forest made a noise if there was no one there to hear it. Were the internal models truly useless, or were they a down payment on better performance in future generations of the Cart?

Further Reading

Cohen, J. 1966. *Human Robots in Myth and Science*. London: George Allen & Unwin.

Dennett, D. C. 1998. *Brainchildren*. Cambridge, Mass.: MIT Press.

Moravec, H. P. 1981. *Robot Rover Visual Navigation*. Ann Arbor: Mich.: UMI Research Press.

Nilsson, N. J., ed. 1984. *Shakey the Robot*. SRI International Center, Technical Note 323.

Walter, W. G. 1950. "An Imitation of Life," *Scientific American* 182(5): 42–45.

———. 1951. "A Machine That Learns." *Scientific American* 185(5): 60–63.

———. 1953. *The Living Brain*. London: Duckworth. Republished in 1961 by Penguin, Harmondsworth, United Kingdom.

Wood, G. 2001. *Living Dolls: A History of the Quest for Mechanical Life*. London: Faber & Faber.

3.   Planetary Ambassadors

B y May of 1992, I had been building my own artificial creatures for almost eight years. They were an artistic success by then, but they had not found an enduring place anywhere beyond my laboratory. I was sure that they could change the way we thought about robotkind, and change our dealings with machines, but I needed a homerun application for them.

So one morning I snuck out of a session of an artificial life conference in Santa Fe, New Mexico, and flew to Los Angeles. I was on my way to meet for the first time one of my teenage heroes. The first man to drive a car on the moon, the commander of *Apollo 15,* and the former chief test pilot at Edwards Air Force Base, David Scott, picked me up in a little old red Toyota. The glory days of the space program it was not. Insurgency it was. By the time we finished lunch, David Scott and I were going to the Moon, or at least our robots were. Now we had a plan for my artificial creatures, a place for them beyond my laboratory.

A Creature Awakening

While I was helping Hans Moravec lug things around at SAIL, I was also working on my own Ph.D. research. The topic was model-based computer vision. I would give my program an a priori three-dimensional model of objects: an airplane was a cylinder with two trapezoidal wings, some cylindrical engines, and a trapezoidal tail. The program then tried to find instances of the models in an image. My programs took hours to run on the same SAIL computer that Hans used for fifteen minutes to analyze each of his snapshots. I graduated and went on a little odyssey of research-scientist positions at Carnegie Mellon University and MIT, and then faculty positions at Stanford and MIT. All this time I worked on model-based approaches to industrial robotics, computer vision, and path planning for mobile robots. When I joined the faculty at MIT as a member of the Artificial Intelligence Laboratory in September 1984, I decided to build my own mobile robot.

Over the first few months I found laboratory space and recruited students. Anita Flynn was a sports-obsessed MIT graduate student who had dropped out of the U.S. Naval Academy when she found to her dismay and outrage that the Navy had a policy against women flying combat jets. It had never occurred to her when she entered the academy that her beloved Navy could have such a grossly dishonorable rule. After a year digging ditches, she was admitted to MIT and completed her undergraduate degree there. Peter Ning, the son of a Taiwanese diplomat, was an electrical engineering undergraduate looking for a fourth-year thesis topic. He was eager to build a microprocessor system—I was a little scared at that prospect. I had started working with Sathya Narayanan, a student, at Stanford and he followed me to MIT. He was a jack-of-all-trades equally at home with software and hardware. However, the cold of his first Cambridge winter did not remind him of his southern Indian home, Madras.

I needed a robot chassis—something that could roll around on the floor. I was not very eager to build something from scratch. Hans Moravec had done that when he had become a postdoc at Carnegie Mellon in 1980. It had taken him years, lots of money, and wasn't very successful. I did not want to repeat that process, and so I tried to see if I could get something ready-made, or hire some of Hans's students from CMU who had successfully built a very simple working system on their second try. Luckily, Anita came across Grinnell More.[1] Grinnell, the son of Trenchard More, one of the twelve attendees at John McCarthy's 1956 Dartmouth workshop, was having disagreements with his high school about how education should be handled. All his energy was channeled into a company that he and two of his friends had started—to build robot chassis. They had developed a robot they called the VECTROBOT that was 45 centimeters in diameter, and had three wheels that steered together and drove together. It was a cylinder that was about 25 centimeters high. I quickly bought one of these from Grinnell. Before long his two partners had left for more stable jobs, and Grinnell started to hang out at our lab.

The robot base took simple velocity and direction commands over a serial line. Now we had to build some sensors on top of it and get some computation to connect the sensors to the actuators. We decided to use as the primary sensor a ring of twelve sonars. These sonars had been developed by Polaroid to measure the distance to an object so that their SX-70 camera could automatically focus. They could measure distances from 30 centimeters out to about 5 meters, with about a centimeter repeatability, although their true accuracy was much less. We also

---

1. Grinnell is now a senior vice president of my company iRobot Corporation. He heads the Real World Interface Division, which builds research robots for labs throughout the world, and develops and markets *urban robots*. One class of the research robots are direct mechanical descendants of the original VECTROBOTS, although the product line has grown considerably. The urban robots are tracked vehicles, small enough to be hurled through a window by a police officer in a terrorism situation or a marine in an urban conflict. The robot can then be sent on a reconnaissance mission, clambering up rubble-filled stairwells, and sending back video of just what is going on in a dangerous building.

wanted to have two cameras on board so that the robot could see its environment. In order for these to have a good view of an office environment, they needed to be mounted at waist height. This meant that the robot had to be higher than tables and desks, so the sonars needed to be mounted at about tabletop height. That way they would have the best chance of sensing tables as obstacles, rather than just seeing the legs and letting the robot decapitate itself by trying to drive through the middle of the table. The result of all this was a cylindrical robot that was about the size of R2D2. Such robots have become common in research laboratories these days, but this was one of the very first. We called it Allen, in honor of Allen Newell at Carnegie Mellon University, one of the founders of the field of AI, and another of the twelve participants at the Dartmouth workshop.

From our experiences seeing and working with Shakey and the Cart, it never occurred to us that we would do the computation on board the robot. That was just not how it was done in artificial intelligence labs, as had previously been demonstrated. The main computers at the MIT Artificial Intelligence Lab in 1984 and 1985 were radically different from what the rest of the world used. There were about twenty machines that had been hand-built in the lab, each with a bit-mapped display (just as any current PC has), a mouse, an Ethernet-like connection to a file server, 128K bytes of main memory, and 16M bytes of virtual memory. And their native language was Lisp, a wonderfully expressive language developed by John McCarthy back in 1959. They were called Lisp Machines. I decided to use a Lisp Machine as the main processor and connect it to the robot via a serial port. I wanted that to be a radio link, but despite valiant attempts by some MIT undergraduates I hired, we were never able to get it working satisfactorily before we scrapped our first robot. Less than twenty years later, such digital radio links are consumer products that people use in the homes every day, but we had to be content with a 20-meter-long cable. On board the robot, Peter Ning, with the help of a longtime engineer at the AI Lab, Noble Larson, built a single-board microprocessor to interface to the external serial line, the serial line to the motors and to the sonar sensors. The onboard microprocessor acted as a switching station shuffling bits to and from the right places—it did not do much computation itself and was largely underutilized.

I had been thinking about how to organize the computation to control the robot. Somehow the system would connect the perceptual processes that processed the raw sensor data to the motor processes that controlled commands to the robot base. The question was how this computational box should be constructed—what computations should it do, how much feedback should go to the perceptual processes and come from the motor processes, and what was the appropriate way to represent and specify its computations. Informally, at least, people thought of this box as the *cognition* box, the heart of thinking and intelligence. The best way to build this box, I decided, was to eliminate it. No cognition. Just sensing and action. That is all I would build, and completely leave out what traditionally was thought of as the *intelligence* of an artificial intelligence.

Just what is intelligence? Dictionary definitions talk about the ability to learn, to understand new situations, or to apply knowledge to manipulate one's environment. Tying these definitions down to objective criteria so that a robot could be evaluated as to whether it possessed intelligence is a daunting and most likely fruitless task. We can make some comparisons, however. A person mostly seems more intelligent than a dog, and a dog more intelligent than a salamander, and a salamander more intelligent than an ant. And perhaps we would be willing to concede that an ant is more intelligent than a wind-up toy. Just what makes one of these creatures more intelligent than another, however, is very hard to quantify.

Judging by the projects chosen in the early days of AI, intelligence was thought to be best characterized as the things that highly educated male scientists found challenging. Projects included having a computer play chess, carry out integration problems that would be found in a college calculus course, prove mathematical theorems, and solve very complicated word algebra problems. The things that children of four or five years could do effortlessly, such as visually distinguishing between a coffee cup and a chair, or walking around on two legs, or finding their way from their bedroom to the living room were not thought of as activities requiring intelligence. Nor were any aesthetic judgments included in the repertoire of intelligence-based skills.

By the eighties most people in AI had realized that these latter problems were very difficult, and over the twenty years since then, many have come to realize that in fact they are much harder than the former set of problems. Seeing, walking, navigating, and aesthetically judging do not usually take explicit thought, or chains of thought-out reasoning. They just happen.

At first blush, my decision to leave out a cognition box seemed to indicate that I was giving up on chess, calculus, and problem solving as a part of intelligence that I wanted to tackle. In fact, this was not my intent. To me it seemed that these sorts of intelligence capabilities are all based on a substrate of the ability to see, walk, navigate, and judge. My belief at the time, and still today, is that they arise from the interaction of perception and action, and that getting these right was the key to more general intelligence.

In hindsight, I can see that there were a number of threads that had come together to bring me to my radical position. During my earlier years as a postdoc at MIT, and as a junior faculty member at Stanford, I had developed a heuristic in carrying out research. I would look at how everyone else was tackling a certain problem and find the core central thing that they all agreed on so much that they never even talked about it. Then I would negate the central implicit belief and see where it led. This often turned out to be quite useful. For a year or so I worked with Tomas Lozano-Perez on algorithms to find collision-free paths for industrial robot arms as they moved through a cluttered work space to manipulate parts, or weld, or spray-paint. I realized that everyone who was working on these algorithms was concentrating on how the obstacles in real space appeared as obstacles in a higher-dimensional mathematical space of motor control—everyone concentrated on how to represent these higher-dimensional obstacles. I decided that instead of representing where the stuff was, I would try representing where the stuff wasn't. In my algorithms there was to be a representation of where it was safe to move the robot arm, and within those constraints they would try to plan a path to get the desired job done. This turned out to provide an immediate payoff and led to some practical algorithms where none had existed before.

In early 1984 while I was still on the faculty at Stanford, Sandy Pentland was holding a seminar series at nearby SRI International. The series was titled "From Pixels to Predicates."[2] The pixels part referred to the

---

2. Ultimately there was a 1986 book published from this seminar series, edited by Pentland. My paper in that volume does not include any of the more radical thoughts I presented that day—rather it is a paper very much in the spirit of the title of the seminar series, based on my earlier Ph.D. thesis.

basic picture elements we see in a digital image on our computer or television screens. Such pictures are made up of a rectangular array of tiny square pixels, each uniformly painted one particular color. Our eyes interpret the juxtaposition of these pixels as corresponding to objects in the real world. The predicate part of Sandy's title referred to the type of representation of the world that Shakey had used. Shakey had used a first-order predicate calculus logic to represent its world. The details of the representation are not important here, beyond recapping that the models Shakey kept of its world were intended to be complete descriptions of the physical layout, so that plans could be made in the model that would be perfectly valid in the corresponding real world. The underlying theme of the seminar series was that computer vision was not doing a good job of taking pixels as inputs and producing predicate descriptions of the world as outputs. Everyone knew that this was the ultimate goal of computer vision, and even the legendary David Marr at the MIT AI Lab, before his untimely 1980 death, had said so explicitly, that the final stage of vision is "an object-centered representation of the three-dimensional structure and of the organization of the viewed shape, together with some description of its surface properties. . . ."[3]

I was invited to talk at this seminar series on the basis of my earlier work. By this time, however, I was getting very frustrated in my work. I was doing little work with actual robots because the computational problems in maintaining detailed representations of the world were so large. It was seeming more and more impossible to me to extract all the information from images or other sensor inputs that the internal-world models seemed to need in order for the computational planners to operate on them appropriately. Real animals on the other hand, and of course Grey Walter's artificial tortoises, did not seem to have enough neural capacity to be doing as many computations as our algorithms demanded. Something seemed to be wrong somewhere. When I gave my talk at Sandy's seminar, I suggested that perhaps perception and action were more directly linked, without detailed world models in between them. I did this through a series of overhead transparencies[4] that removed the

3. See p. 38 of Marr's 1982 posthumous book *Vision*, listed in "Further Reading" at the end of this chapter.
4. Overhead transparencies were what we used to use in academia before the advent of PowerPoint and LCD projectors.

cognition box that had traditionally linked perception and action, the box that everyone had implicitly assumed just had to be there. This was the central belief that I negated. I made perception and action bigger boxes, based on what was known about how much of animal brains are devoted to them, and made them overlap, without any cognition box at all. I put cognition in a little cartoon cloud representing the thoughts of an external observer of the complete system of the robot—world, perception, and action.

At the time that I gave the talk, I could not see through the murk well enough to envision how such a system would actually be built. Later, at MIT, while the physical Allen was being constructed, I did work out a way to do it. It seemed to me that insects could do much more than any existing mobile robot at the time. They could move at speeds of a meter or more per second while avoiding obstacles, evading predators, and finding food and mates. They had only some tens of thousands or perhaps a few hundred thousand neurons, and each of those computed very slowly when compared to a digital computer. Multiplying out by the number of neurons still said that an insect was not very far off in computational capacity from a good-sized digital computer. In theory at least, we should have been able to build a control system for a robot, using a digital computer, that would get similar levels of performance. What was the key about the way in which the nervous systems of insects were organized that let them perform so well with so little computation?

During the summer of 1985, I was stuck for a few weeks in a house on stilts in a river in southern Thailand, confined to the house by well-meaning in-laws, worried about my safety in the neighborhood. While everyone else around me was busy speaking Thai, I had hours and hours per day to just sit and think. I started out by drawing conventional diagrams of how the computation should be organized. Boxes with arrows. Each box represented a computation that had to be done, and each arrow represented a flow of information, output from one box, directed as inputs to one or more other boxes. Such box and pointer diagrams were, and still are, common. Each box schematically represented a very large-scale computation that might take seconds or more on the main computer. When one looked at these diagrams, with multiple boxes on any path between sensors and actuators, it became clear that the system was going to be very slow and not very reactive to the environment in the way that insects are. Slowly the idea dawned on me. Make the computations simpler and simpler, so that eventually what needed to happen in each box would take only a few milliseconds (i.e., a few thousandths of a second). But rather than compensate by building thousands of boxes, I realized that I could get by with just a handful or two of simple boxes, the same number, roughly, as in the original diagrams that I had drawn. The key was that by getting the robot to react to its sensors so quickly, it did not need to construct and maintain a detailed computational world model inside itself. It could simply refer to the actual world via its sensors when it needed to see what was happening. This was the second use of my research heuristic. If the building and maintaining of the internal world model was hard and a computational drain, then get rid of that internal model. Every other robot had had one. But it was not clear that insects had them, so why did our robots necessarily need them.

How to organize these computation boxes in a general way was still a problem. Again, thinking about insects gave an insight. Insects, and indeed all creatures, evolved over time. They neither started out fully formed with all their capabilities, nor did they remain static over hundreds or thousands of generations. In the main, creatures started out with simple capabilities and, over time, developed more sophisticated capabilities. Of course in some circumstances abilities withered away when they were no longer needed—we humans no longer need to digest grass, and our appendix has shriveled in size, so that now it is a mere appendix to our digestive system. But in general, complex capabilities are built on top of simpler capabilities, often through the addition of new neurons.

This was the metaphor I chose for my robots. I would build simple control systems for simple behavior. Then I would add extra control systems for more complex behavior, leaving the older control systems in place, still operating. If necessary, the newer control systems might occasionally subsume the capabilities of the older system when they knew better how the robot should act. And so the layers would be added one after the other, emulating the historical process of evolution of more and more complex neural systems in real animals.

For Allen I targeted three layers. The first was a control system that was to make sure that the robot avoided contact with objects whether they be stationary or moving. That meant it had to shy away from objects detected by its sonars no matter whether it was trying to move in some direction, or if somebody or thing came close to it. The second layer would give the robot wanderlust, so that it would move about aimlessly. Because of the existing lower-level control system that avoided contact with objects, the aimless wanderer could be very simple without having to worry about collisions. The third layer was to explore the world purposefully whenever it perceived anything interesting in the distance—it would head over in that direction. Again, this third level did not have to worry about avoiding collisions, for the lowest, most primitive layer of control took care of that. And, if the exploration layer could not perceive anything interesting to go and explore, the aimless-wandering layer would have control and let the robot browse the world until the upper layer did find something of interest.

This was the architecture (called the subsumption architecture) for building my artificial creatures. Their nervous systems were expressed as circuit diagrams connecting a dozen or so simple computational elements to input sensors and output actuators. The circuit elements treated their input wires as continuous sources of signals and generated similar continuous signals on their output wires. On the Lisp Machine we were using as our main computer, I quickly implemented a circuit simulator that acted as though each of those simple computational boxes were running continuously and in parallel. That was rather easy in fact. I spent more time building a tool to lay out a nice graphics picture of the circuit diagram, and another tool that monitored how much computer time it took to simulate each individual circuit element and provide a real-time graphics display of it. The former was not much use, and the latter showed that it took almost no time at all to simulate the whole circuit. Now I just needed to connect everything to a real robot. But the robot itself was not yet ready despite the hard work of Peter, Anita, and Sathya.

I decided to implement a simulator of the actual robot, including some simulated physics of how the sonars would perform. The control system that I had built for the real robot could be used to control the simulated robot. This is a common technique today but was not so common at that time. Soon I had my control system debugged. In the simulated world it was able to drive the robot around without a collision with any obstacles. It was able to maintain a simulated velocity of almost half a meter per second, operating continuously, with no need for the simulated robot to stop and ponder at any point. This simulated robot was much better than either the Cart or Hilare. But much better than that, I could put simulated moving obstacles in the world and my simulated robot would skirt right around them as it moved through simulated space with real wall clock time ticking away. This was something that no other existing robot could do—operate in a dynamically changing environment. And why could my simulated robot handle it? Because it was using the world as its own model. It never referred to an internal description of the world that would quickly get out of date if anything in the real world moved.

In October of 1985 the Second International Symposium of Robotics Research was to be held in Gouvieux-Chantilly, France, just north of Paris. The first such symposium had been put together by Michael Brady from the MIT AI Lab in 1983, and had been held at Bretton Woods in New Hampshire. The idea was to bring together the fifty most innovative researchers in robotics from around the world for an intensive week of presentations and discussions. I had been invited to the first for my work on industrial robots at MIT when I had been there as a research scientist from 1981 to 1983. On the strength of that earlier work I was invited back again in 1985. I desperately wanted to get Allen working in time for the conference so that I could show how new and differently it would work. By the time the papers were due, it was not yet working, so I could only include some simulation results. I did manage to get Allen just barely working before flying to France, and had a poor-quality video of its first few runs. It moved down corridors at a brisk pace, centering itself between the walls. It diverted itself around oncoming pedestrians. It could be herded by a group of people surrounding it on three sides so that it only had one direction to go without causing a collision. It was glorious.

At the conference, when it came my turn to present, I got up and explained how I had built an intelligent robot by denying the main tenet of artificial intelligence. My system eliminated any process of reasoning, or going through chains of thought. Instead, it was completely based on unthinking activity, the direct connection of perception to action. And I claimed that my robot worked better than anything built before using the old approach. Without quite being explicit I was saying that all previous work in artificial intelligence was misguided and that my new approach would change everything.

Georges Giralt, of Hilare fame, the chair of the conference, and Ruzena Bajscy, now head of computer science at the National Science Foundation but then head of the robotics laboratory at the University of Pennsylvania, were sitting together at the back of the room. Years later Ruzena told me that she and Georges were whispering to each other during my presentation: "Why is this young man throwing away his career?" In the video my robot did much more than anything any previous robot had done, if one just looked at it as an external observer. But it did it in too simple a way! The audience could not accept that this was serious work. It did many more things much faster than any previous robot. But there were not reams of equations, nor multipage complex algorithms implemented on the robot. This could not be serious robotics research. It was this talk that split me off from the mainstream of robotics research. The arguments that started that day continue to today. That talk, that day, was the defining point of my academic career. My students and I face the same arguments today as we demonstrate our humanoid robots interacting with people in humanlike ways, but with behavior that is generated by relatively simple rules, built on top of computationally intensive perceptual processes.

I returned to MIT more excited than ever about the direction of my work. I had tasted battle, and it was fun. I knew that for my tenure case I would need to publish my new work in journals. There was a new IEEE journal called *Robotics and Automation,* so I sent an updated version of my Gouvieux-Chantilly paper to the editor, George Bekey of the University of Southern California. He sent it out to three anonymous reviewers, all of whom recommended its rejection. The work was too new, too untested, and too simple. George accepted the paper anyway, and it was published in April 1986. It has become one of the most referenced papers[5] in all of robotics and computer science.

A new Ph.D. student joined our group—Jonathan Connell. Initially he added more layers of control to Allen so that it was able to follow walls, align itself through doorways, and very methodically explore

5. It is reprinted in my book *Cambrian Intelligence,* referenced at the end of this chapter.

office areas. But soon we reached the limitations of the sonar sensors that we had on board Allen and wanted a robot that could manipulate the world in some way, not just wander around in it.

We started work on a new robot, Herbert, named for Herbert Simon, Allen Newell's colleague at Carnegie Mellon. This robot was to have all its processing on board, using simple 8-bit processors, a laser scanner to get three-dimensional data, and an arm that could pick things up from tabletops and the floor. Herbert's task was to find empty soda cans around the lab and collect them. It produced much more controversy over its lifetime than Allen ever did, but its most important contribution for the thread we are following here is that it paved the way technically for my most successful robot, Genghis. By building Herbert we became the first mobile robot group to have more than one robot. As with kids, once you have broken the taboo of having a second one, it becomes easier and easier to continue having more. It also gave us experience putting computation on board a robot.

A Robot That Walks

By early 1988 our research group was working on multiple robots. One had an arm, one had vision, one had a laser scanner, and two could chase each other around. But all of them rolled around on wheels. None of them walked on legs like the vast majority of land creatures do. There were a few walking robots being built, most successfully by Marc Raibert, who was also at the MIT AI Lab. His robots hopped on one or more extendable legs. Marc's robots had to keep moving quickly in order to maintain their balance. A handful of robots had been built with articulated legs, but all of them took seconds or more to take each step. All of the robots, hopping or otherwise, walked or ran in a straight line, with no other purpose than to demonstrate that they could walk.

I wondered whether we could build a walking robot that did more than that. But first I needed an implicit assumption to negate. What was it that every one was assuming, that would make things easier by being

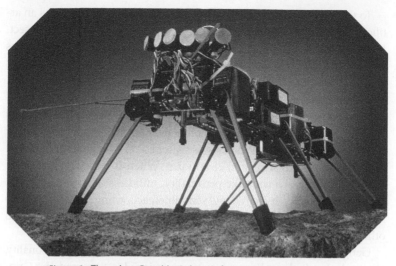

Figure 1. The robot Genghis. It has six legs, making it look very insectlike. The six pyroelectric sensors arrayed at the front of the robot allow it to sense the presence of heat-emitting mammals.

negated? While watching videos of insects walking over rough terrain, I noticed that many times they would miss a foot placement and stumble. This was the key. Everyone had been assuming that walking robots needed to stay balanced. What if we let them fall down, let them scramble more than walk, just the way insects operate?

Grinnell More, spending a day a week in our lab, quickly mocked up a two-legged walker that dragged its tail along the ground. We operated it with a model-airplane controller and found that we could make it scramble easily over quite difficult surfaces. Colin Angle[6] was a new undergraduate in my lab. He, Grinnell, and I decided to build a six-legged version, with three identical legs symmetrically arranged on each side of a straight spine. Grinnell built the mechanical robot, Colin built a four-processor onboard computer system, and I programmed it. Eventually we named the robot Genghis (figure 1) because it was able to walk over anything in its path as it followed a person, attracted by the usually invisible infrared glow that the warm body of every mammal emits.

6. Colin was a cofounder, and is now CEO, of my company iRobot Corporation.

To this day Genghis[7] has been my most satisfying robot. It was an artificial creature. It looked like a six-legged insect. A *big* six-legged insect. It walked like a slightly ungainly stick insect. Its only user interface was an ON-OFF switch. When powered off, it sat on the floor with its legs sprawled out flat. When it was switched on, it would stand up and wait to see some moving infrared source. As soon as its beady array of six sensors caught sight of something, it was off. As long as it could track its prey, it kept going, ruthlessly scrambling over anything in its path, solely directed toward its goal. When it was switched off, Genghis was a lifeless collection of metal, wire, and electronics. When it was switched on, it came to life! It had a wasplike personality: mindless determination. But it had a personality. It chased and scrambled according to its will, not to the whim of a human controller. It acted like a creature, and to me and others who saw it, it felt like a creature. It was an artificial creature.

Genghis's physical form undoubtedly gave it some of its personality, although that had not been a conscious design decision made by Grinnell, Colin, or me. More importantly, its software gave it "life." It acted in ways that appeared lifelike—it acted as a creature might act in similar circumstances.

Of course, software is not lifelike itself. But software organized the right way can give rise to lifelike behavior—it can cross the boundary from machinelike, which is how we normally think of software, to animal-like. The appendix at the end of this book completely specifies the software that made Genghis operate. In this chapter we will examine just a little of its software to demonstrate how the transition from machinelike to animal-like can occur.

The software for Genghis was not organized as a single program but rather as fifty-one little tiny parallel programs. Building on some conventions from computer science, we called these programs *augmented finite-state machines,* or AFSMs.

In our case each of these AFSMs was similiar in complexity to the software that runs a soda machine, something that could be in one of just a couple of states and that could store one or two 3-digit numbers. The program that runs a soda machine is either in the state that there is

7. Genghis has resided in the National Air and Space Museum in Washington, D.C., since the mid-nineties.

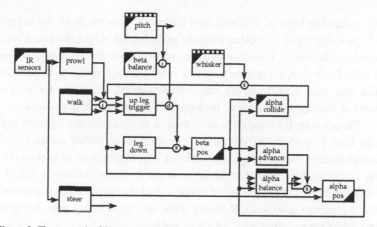

Figure 2. These are the fifty-one augmented finite-state machines (AFSMs) that transform Genghis from a pile of metal, plastic, and silicon, into an artificial creature. Each AFSM with a solid black bar as its top is unique. Each one with a uniform outline is actually replicated six times, once for each leg. Each one with a striped top is replicated twice, either for front-back symmetry or for left-right symmetry. AFSMs with a triangle in their upper-left corner receive inputs from one or more sensors. Those with a triangle in the lower-right corner have outputs that command a motor. Where three outputs or three inputs appear, there really should be six, which fan in, or fan out, to the six copies of the local circuit applied to each leg.

enough money deposited to pay for a soda, or there is not yet enough. In either state the machine will accept more coins and keep a running total of what has been deposited as a single number—$1.25, say. If it is in the *not enough* state, then it ignores pushes of the soda selection buttons— e.g., the GINGER ALE button. If it is in the *enough* state, then when a soda selection button is pushed, it opens up the control door for the corresponding soda, subtracts the cost of the soda from its running total, and opens coin-control paths to return the correct amount of change. It immediately changes to the *not enough* state, and resets the running total to zero. There is nothing else that the soda machine control program can do.

Figure 2 shows the network of fifty-one AFSMs that controlled Genghis. None of them is more complex than the program to control a soda machine that we have just outlined. The only difference is that these programs, or AFSMs, can send numbers to each other on fixed wires. The AFSMs were the boxes in my box and pointer diagrams I had been drawing back in Thailand, and the paths, or wires, from the output of one to the input of another were the pointers. We might imagine a soda machine that sends a message to the company headquarters, perhaps over a cell-phone link, every time someone buys a soda, telling the home base that one more ginger ale, say, has been dispensed. These are the simple sorts of messages that get sent between AFSMs in Genghis's network.

The software for Genghis was written in an evolutionary manner, and the first forty-eight AFSMs allow Genghis to scramble around over rough terrain, mostly keeping its balance and responding to obstacles in its way. The most important thing to know for what follows is that the *walk* machine, toward the left of figure 2, had six outputs that sequenced the six legs to take steps. Without those messages being sent, Genghis would just stand there and not try to walk.

The last layer of behavior was provided by the three AFSMs named *IR sensors, prowl,* and *steer.* This layer turned Genghis into a predator. It would lie in wait for a heat-emitting source to come into view, lock onto that source, and chase it over whatever terrain was in front of it. These, of course, are the descriptive terms that an observer might use when seeing Genghis in action. As the reader can by now guess, the implementation of these behaviors was done with much less explicitly intentional components.

The *IR sensors* machine received input from the six pyroelectric sensors that are clearly visible at the front of Genghis in figure 1. Each one of those was OFF if it could detect nothing in the direction it was pointed, and ON if it could see a change in infrared radiation. These sensors were identical to those used as motion sensors to turn on outside lights on people's homes, and were tuned to just the radiation band that is emitted by all mammals from the heat of their bodies. Thus these sensors were very good at detecting a moving person in their line of sight. A slightly formal definition of the *IR sensors* machine is as follows:

> *IR sensors.* Continuously output a list[8] of which of the six pyroelectric sensors have triggered ON within the previous half a second.

8. In reality, the list was represented by a number in the range 0 to 63, where each of its six bits represented in binary corresponded to the state of one of the sensors.

The output of the IR *sensors* machine was fed as an input to two other machines. The first was the *prowl* machine.

> *prowl.* Input: *pyroelectric sensor list.* Continuously output an inhibitory message, but if ever any sensors are listed on the input, then stop doing so for five seconds.

The output of the *prowl* machine was connected to inhibit the outputs of the *walk* finite-state machine. So the combined effect of our two new machines was to make Genghis stop in its tracks if it saw no infrared activity ahead of it.

When there was infrared activity in the sight of the forward-looking pyroelectric sensors, then the robot started to walk for a few seconds and would continue to do so if it maintained a view of infrared activities. Now note the way that the configuration of the robot interacted with the control machines to influence its ultimate behavior. The pyroelectric sensors pointed forward, and the robot walked forward. So if the robot saw some infrared activity, it walked *toward* that activity. To an external observer it appeared to see the activity in front of it and willfully walk toward that activity. If the sensors had been rotated to point to the back of the robot, however, Genghis would have walked *away* from the infrared activity. Genghis had no internal notion of toward, or forward, or away. That was all embedded in the interactions of its sensors and actuators, mediated by the very simple, mindless AFSMs that we built for it.

Another machine, *steer,* also received the output of the IR *sensors* machine.

> *steer.* Input: *pyroelectric sensor list.* Continuously count the number of left sensors and right sensors that are ON. If there are more on the left, then send a message to the left legs to take smaller steps, and if there are more on the right then make the right legs take smaller steps.

Think for a moment about the effect of taking smaller steps on one side of Genghis's body. If you stand and take normal-sized steps with your left foot, but small steps with your right, your body will turn to the right. The same thing works for Genghis. Thus the machine *steer* was able to steer Genghis toward any source of infrared. Since Genghis walked only if it could see such a source, the combined result was that it would walk toward any source of infrared, following it if it moved. But it did not even know whether the source was moving. It would just follow it.

Now Genghis would sit and wait for some moving source of infrared to come along and pass in front of it. Then Genghis would start walking toward it, steering left and right to keep the target in sight, scrambling over any obstacles in its way.

The first forty-eight AFSMs of Genghis were just as mindless as these last three. You can read all the details in the Appendix should you choose. The software that controlled Genghis was very simple. It was a collection of extremely simple, local computations, which when placed together and in the context of the physical robot, brought that inanimate object to life. It transcended the boundary between living and nonliving. The software itself was certainly not profound. It was rather straightforward, in fact. The software's behavior, however, was profound.

Notice that the high-level observable behaviors of Genghis were not reflected directly in computations carried out in the collection of finite-state machines that made it behave like Genghis. There was no place that represented Genghis's prey and its prey's trajectory. There was no place that represented the lay of the land out in front of Genghis, over which it must scramble. Further, there was no place inside the control systems of Genghis that represented any intent to follow something, or any goal to reach it. However, to an external observer they were the easiest ways to describe Genghis's behavior.

There is a deep philosophical question lurking here. If Genghis did not have its intentions represented anywhere, then did it really have intentions? Or did it just appear to have intentions? These questions beg the further question of what it would mean to have the intention actually represented. Would it, perhaps, be necessary to have a computational homunculus, or perhaps *inseculus,* inside of Genghis in order for it to legitimately have intentions? But even so, there would remain the reductive problem of what should the structure of that internal model be, and would it in turn need its own little homunculus inside it. Turning the questions around another way, we could examine the computational circuits of real insects. When we do, we find similar, although somewhat differently organized, neural circuits that, again, do not have explicit representations of intention. So does a real insect have intentions, or is that overanthropomorphizing it? If we say that it does not have intentions, then we can climb up the so-called evolutionary ladder and ask similar questions about reptiles, mammals, apes, and finally humans. Hardly any philosopher will deny the intentionality of humans, but others may deny intentionality at various points in the nonhuman spectrum.

Although Genghis captured animal-like behavior, philosophers, both professional and those of the just-plain-folks ilk, may have deep troubles in assigning it intentions.

We will return to this set of questions in many other formats throughout the rest of the book.

Situated and Embodied Robots

The insectlike Genghis robot was a turning point in the life of my research group. It was the beginning of our own personal Cambrian explosion. We quickly generated tens of robots with different mechanisms and different capabilities, but all were based on the central ideas that Genghis demonstrates. We were in Cambridge, and our robots took on all sorts of fundamentally different body plans, radiating out from our initial set of design ideas. Surely these two facts justify the Cambrian pun.[9]

All of our robots were based on two fundamental principles: *situatedness* and *embodiment*. These two terms can be a little fuzzy at times, but I like to use the following definitions for them:

A *situated* creature or robot is one that is embedded in the world, and which does not deal with abstract

9. I am indebted to Tony Prescott for the original linguistic suggestion.

descriptions, but through its sensors with the here and now of the world, which directly influences the behavior of the creature.

An *embodied* creature or robot is one that has a physical body and experiences the world, at least in part, directly through the influence of the world on that body. A more specialized type of embodiment occurs when the full extent of the creature is contained within that body.

Under these definitions an airline reservation system is situated but it is not embodied. A robot that mindlessly goes through the same spray-painting pattern minute after minute is embodied but not situated.

Our work with Genghis had shown us the importance of these two principles. A traditional walking robot, such as the Ambler built at Carnegie Mellon, started out by scanning the world in front of it to get a full three-dimensional model. It then planned a nominal path for itself between the obstacles. Then it planned where to put each foot so that its body would follow that path. Then it planned out the required torques on each of its joints to get its feet to the right places so that its body would follow the desired path. Finally it moved its feet. Genghis, on the other hand, started by moving its feet. Its embodiment, its physical being, coupled its six legs together, so as each of them went about its own pattern of activity, the complete body of Genghis followed an emergent trajectory that was a product both of its actions and its situation in the terrain of the world. The *chase prey* layer of control situated in the world modulated the lower-level behaviors of the individual legs so that Genghis could carry out a task *in* the world, following prey, without ever having to plan ahead its trajectory or its every move.

With these lessons learned, we were able to fill out the behaviors of the robots of our Cambrian radiation.

In programming Herbert, Jonathan Connell deliberately chose to legislate that it could not maintain any internal memory for longer than three seconds. Nevertheless, it was able to wander around the lab, essentially treating it as a maze, and keep its virtual "left hand" on a wall so that it did not get lost. It could find a soda can using its laser scanner, pick it up and drop it again if it was full, and return any empty soda can it found to the place it was switched on. Some people got confused by this work, thinking that we were claiming that no animals, even humans, needed to retain any memory longer than three seconds. Of course, that was not what we were claiming; rather, we were seeing just how far we could push the idea of no central place for cognition, leaving all long-term state out in the world: "There is nothing in my hand, therefore I must be looking for a soda can" versus "There is something in my hand, therefore I must be trying to return it home." Herbert was the ultimate situated robot.

My student Maja Mataric built a robot, Toto, on a smaller version of Grinnell More's wheeled bases, that navigated around using a compass and sonar and built maps of its environment without any conventional data structures. Instead, the AFSMs, similar to those that Genghis used, modified themselves and, in ensembles, essentially came to represent the office environments that Toto encountered on its travels. Maja, too, pushed on the lack of a central cognitive engine. Toto acted just like a robot that had a conventional map of the world, except that it was all contained in the low-level sensory and motor systems. The map, rather than being explicit, was a modulation of how Toto was situated in the world—it determined Toto's every reaction to the world through its embodiment, and so Toto acted as though it had a map that it had built itself. There was no separation, as in conventional software, between the part of the system that built the map and the part that used the map. They were one and the same thing. This goes against all the basic tenets of computationalism, and was very hard for most traditional AI researchers to appreciate.

Another student, Ian Horswill, started out by experimenting with simple vision systems for Allen and ended up building yet another robot based on Grinnell's smaller-wheeled bases. This robot, Polly, was designed to be a tour guide for visitors to the MIT AI Lab. It patrolled the corridors, using visual processing to follow them and avoid any obstacles. Polly had to somehow detect visitors to the lab in order to give them tours.

Here is where an analysis of its ecological niche, its situation in the world, became important. All the graduate students in the lab were quite sick of Polly wandering the corridors, prattling on, using a cheap voice synthesizer. When they saw it coming along the corridor, they would simply give it a moderately wide berth so that they could get to where they were going as quickly as possible.

Visitors on the other hand, we reasoned, would be excited to see a robot wandering the corridors and would most likely stop to look at it. So Polly treated anything that was in the middle of a corridor, and tall with vertical edges for the first fifty centimeters from the ground, as a potential visitor. It would stop, verbally offer to give a tour, and wait for a positive response. Since its only sensor was its camera, it would verbally suggest that in order to answer in the affirmative, the visitor should shake one of their feet. On detecting the motion, Polly knew it had a customer. Packing crates or other debris temporarily left in a corridor never answered affirmatively, and Polly would soon wander off looking for another candidate.

Polly operated successfully for a few months, running for about two hours every afternoon. Once it had a victim, it would wander around the seventh floor of our building in a purposeful way, trying to visit every place that it knew about. It did not have an accurate metrical map, rather a hand-derived map similar to the ones that Toto could acquire by itself. It used vision to match to previously recorded images at the tour highlights, and would announce such interesting things as "This is the coffee room and kitchen," as it navigated through a fairly standard office environment. Polly did manage to give a few spontaneous tours to genuine visitors to the lab, and it did it with a camera held on by Velcro, oscillating back and forth twenty degrees as it rumbled along the corridors. We wanted to demonstrate that Polly was surely not building accurate internal models the way Shakey had tried to do—it was swimming in a sea of uncertainty.

Colin Angle and Cynthia Breazeal collaborated on a much more ambitious series of six-legged walking robots. Colin built twin robots Attila and Hannibal that each had nineteen motors, eleven onboard computers, and over a hundred sensors. They really were too complex for us at the time, but Cynthia gamely set about writing AFSMs for them in a slightly higher-level language that I had developed, called the Behavior Language. Eventually she produced over 1,500 AFSMs that let the robots do all sorts of things. They were robust to sensor failure, and, through a model of pain induced by inconsistent sensor readings, they were able to ignore bad sensors and later reintegrate them if they started operating correctly again. Cynthia implemented a couple of models of gait production from the literature on how insects walked, and uncovered an inconsistency in one of those models. Rather than having explicitly sequenced leg motions as in the case of Genghis, above, Attila and Hannibal were dynamic in the gait they generated, it being an emergent property of what the individual legs were encountering and the messages they were sending to each other. Furthermore, the legs were able to cooperate when the robots encountered rough terrain, lifting the body together, holding things up while one leg searched for a difficult foothold, and backing up and going around obstacles as necessary. At 1,500 AFSMs we felt that we had shown that our approach was scalable.

More and more robots continued to sprout at our lab. They included Bodicea, a pneumatic six-legged walker designed for the surface of Mars, using the ambient carbon dioxide as the working fluid; the Nerd Herd of twenty identical robots[10] that cooperated without explicit communication much as social insects do; a series of small robots designed to move lunar regolith around to build shielding for human habitation structures; and yet more insectlike cooperating robots. The common thread of all these robots was that, like Genghis, they were artificial creatures that had their own independent agendas in the world. They were built from layered control systems with no central cognition boxes and they coupled sensors to actuators, much as insects do, through very short neural pathways. They were situated and embodied.

Getting to Mars

While work was going on in my research group at the MIT Artificial Intelligence Laboratory, we were also active in trying to get one or more of our robots into space, and onto another planetary surface. At the same time as Colin Angle and I had been building Genghis, Anita Flynn and I had been speculating about how these sorts of robots might make space exploration cheaper.

10. They were all very similar in that most of them did not work most of the time.

In the late 1980s, JPL (the Jet Propulsion Laboratory in Pasadena, California) had set out a plan to send a mobile robot to Mars. They had built a new rover beyond the one that they had in the late seventies. The new rover, named Robbie, weighed more than a ton and moved at about 1 cm per second under visual control. The baseline cost for the mission being discussed was $12 billion. In the political and budgetary climate of the time this seemed a very unlikely prospect.

Anita and I were buoyed by the rapid success that Colin and I had had in building Genghis. We wrote a paper, titled "Fast, Cheap, and Out of Control: A Robot Invasion of the Solar System," which appeared in the *Journal of the British Interplanetary Society* in 1989. Our idea was that rather than sending one 1,000-kilogram robot to Mars, or any other place for that matter, we would be better off sending one hundred 1-kilogram rovers. It seemed to us that a small rover could accomplish much of what a large rover could do. By sending only one-tenth of the payload, the cost of the mission would go down drastically, as launch mass cost was a large part of the total mission cost. That was the cheap part. By developing a small rover rather than a large one, the development time would go way down. That was the fast part. Furthermore, with the redundancy of many rovers, ground controllers would be much more likely to send a sacrificial rover off to explore an interesting but dangerous locale. Lastly, by making the rovers intelligent, as intelligent as Genghis, say, there would not be as much need for constant ground control supervision, and the robots could be allowed to run autonomously. That was the out-of-control part. It also contributed to making missions cheaper.

When Anita and I first presented these ideas at a workshop at JPL in the summer of 1988, right in the middle of Genghis's construction, our welcome was not very warm. One engineer got up and lambasted the idea, saying that scientists waited fifteen years to get an instrument flying on a mission, and they "wanted a big instrument, not a little bitty one." We were rather discouraged, and perhaps that disappointment contributed to the defiance of our title when we published our paper. Now I admit "Fast, Cheap, and Out of Control" was a little exuberant, but many people did pick up on it.[11] We were quite excited about the

11. "Fast, Cheap, and Out of Control" became something of an underground slogan for many on the Internet, expressing the grass roots explosive growth of that medium, and

ideas and just needed to find someone with a rocket to get us to another planet.

David Miller and Rajiv Desai, two young Ph.D.'s at JPL, quite liked our ideas and invited Colin Angle to come spend the summer of 1989 working with them. While there, Colin surreptitiously built Tooth, a half-kilogram microrover, using the software system we had developed for Genghis. The building of Tooth had to be clandestinely because Colin was in a software group at JPL. Soldering irons were considered too dangerous to put in the hands of programmers and were forbidden. It was also difficult for David and Rajiv to requisition the parts necessary to construct Tooth. So in a strange turn of events for a research university like MIT, I ended up donating parts (very unofficially) to NASA, a traditional sponsor, so that Tooth could be constructed. Tooth had four wheels and was able to navigate about and pick up small rocks with its gripper. Its performance was good enough to convince Donna Shirley, the JPL Mars mission manager, that perhaps there was some future to small rovers, and she set up a tiny program for Miller and Desai to continue the work after Colin returned to MIT.

Rajiv Desai and David Miller had soon built their own microrover. It used the same software architecture as Tooth and at first even used some of the same code, but it had a much better mechanical design. It was a six-wheeled device known as a rocker-bogey, designed at JPL, and very good at climbing over relatively large rocks. The first robot was called *Rocky* in honor of this mechanism, and so naturally subsequent generations were called *Rocky 2, Rocky 3*, etc. But the projects did not get much respect at JPL and in fact provoked some outright resentment. Building small robots on a shoestring budget was not a good way of attracting large sums of money from NASA headquarters to keep the large infrastructure of JPL afloat. In fact, there was a danger that it might lead to cuts in budgets.

Colin Angle, back at MIT, and I were depressed. We did not see how the JPL connection would get our robots to other planets. Before long we teamed up with Bruce Bullock, the chairman of an artificial intelligence

---

Kevin Kelly used part of it for the title of his 1994 book on the new modes of thinking in artificial intelligence. Errol Morris used it as the title of his 1997 general release movie that featured me and three other misfits (a lion tamer, a topiary gardener, and a keeper of naked mole rats).

government-contract research firm ISX, located in the Los Angeles area. We started a new, for-profit company named IS Robotics,[12] whose explicit mission was to engage in commercial exploration of the Moon and of other planets. We put together a somewhat optimistic business plan, selling scientific-instrument time to scientists via their National Science Foundation grants—hardly a surefire source of dollars. We went to many small private launch companies, none of whom had a successful launch at the time, looking for cheap deals on getting to the Moon as our first target. We got some Hollywood insiders involved in the hope they would be able to sell movie rights ahead of time and in fact simultaneously produce the movie of the epic. We talked to cartoon series writers about a cartoon based on the characters of the robots that would fly to the Moon. We put little athletic shoes on the feet of six-legged robots and hawked the idea of footprints on the Moon as the centerpiece of an advertising campaign. We just needed a lousy $22 million and a lot of luck, and everything would work out right.

We were struggling and at a nadir when Bruce met David Scott. He had already been to the Moon, liked our technical approach, and wanted to do something new in space that was memorable, by being one of the first to commercialize space activity. My trip to Los Angeles in May 1992 was to convince Dave that we had a plan that might help him achieve his goals.

Dave realized that the perfect people to help us out were the Ballistic Missile Defense Organization. BMDO at this time was the official name of what was better known in the popular press as Star Wars, the ballistic missile defense shield conceived of by Edward Teller and Lowell Wood, and promoted by Ronald Reagan. With the collapse of the evil empire of the Soviet Union, BMDO was in the political wilderness. Their reason for existence was in question. Their unofficial strategy was to compete against NASA and show that they were a better space-faring organization.

At this time in 1992, BMDO had two things going for it that they thought of separately, but for our purpose, when combined, were perfect. The first was a piece of technology. They had kinetic kill vehicles. These were small spacecraft that had flown in space, weighing 23 kilo-

12. IS Robotics has now metamorphed into iRobot Corporation.

grams, fully fueled. The intent was to put up hundreds or thousands of these "brilliant pebbles," which would sit and watch for unannounced ballistic missile plumes using a suite of onboard sensors. With approval from ground control they would home in on the missile as it climbed through the atmosphere and smash into it kinetically (i.e., just using its own momentum rather than any explosives). These kill vehicles were incredibly maneuverable and in fact were tested on the ground by having them hover above a net in a test facility. Fully fueled, they would be able to leave lunar orbit, should one find itself there, and soft-land on the surface of the Moon. Without the need to carry their ballistic-missile–seeking sensors, they had plenty of payload capability to carry one of our small robots.

The second thing BMDO had was a plan to put a satellite in orbit around the Moon. Nominally this was to test out their space capabilities, but certainly one of the unspoken subtexts was to show that BMDO could get something into orbit around another planetary body— something NASA had not done for over ten years. Dave noted that their launcher and satellite could fit a 23-kilogram package on board, a kinetic kill vehicle carrying one of our robots.

Dave Scott managed to convince some of the folks at BMDO that it would be a real coup for them if they could do a soft landing of a robot with pinpoint accuracy next to the remnants of the *Apollo 15* lunar excursion module. We could wrap it in the veil of science by doing some metallurgical studies of the legs of the LEM, a known metal that had been the subject of twenty years of exposure to the lunar environment.

Once we got the go-ahead, we quickly built six different prototype six-legged walkers over a six-week period. "Fast and cheap" was working! The rovers had to weigh only half a kilogram, and they required a new onboard computer architecture. At the end of the summer of 1992 we had three good test rovers. By this time Colin Angle and I had been joined at our company by Helen Greiner, another former Artificial Intelligence Lab student. Helen did mechanical design, Colin electrical, and I wrote the software. It was intense work but fantastically satisfying. We were getting ready to fly an Earth-based test. But then, even at BMDO, bureaucracy was starting to set in, and it was not until October 1993 that the first real test happened.

It was at Edwards Air Force Base, up on the hill in the same building where the *Saturn V* engines had been tested for the manned Moon missions. Our flight vehicle was installed in the payload bay of a kinetic kill vehicle and left there for some days, with no communications, to simulate the time it takes to get to lunar orbit. The kill vehicle itself had been modified with landing legs. Everyone was evacuated from the test site, lest the rocket motors blow up. A countdown happened. Just like a real one. The kill vehicle lifted off, and hovered in the Earth's gravity, six times what it would encounter on the Moon. It automatically flew over to a mocked-up lunar surface, then descended and landed with only a minor thump. The first major hurdle of the mission had gone without a hitch.

Our robot was encased in a pod that was weighted so that no matter which way it fell onto the surface it would passively roll to be right side up. We had dropped this pod with our robot in it onto concrete from a meter up, over one hundred times, to test it out. The kinetic kill vehicle shut down its engines and delivered a kick to the pod, ejecting it onto the hard surface from a height of about 50 centimeters. It landed right side up and did not even need to roll. Second hurdle passed. A cheer rose up in the control room. Now the robot had to perform.

The robot had its legs all folded up to minimize its volume so that it could fit into the cocoon. To get out it had to use one of its folded legs to unlatch a retainer holding the pod together. The robot realized that its mission had started right on cue, again with no explicit control from the mission overseers. "Out of control" was working. The pod opened to thunderous cheers from the control room. Third hurdle passed.

Grendel, the robot, untangled its legs. It stood. It started walking away from the lander, looking for a place to scoop up some soil with its underbelly shovel. The control room crowd went wild. This scheme actually worked!

Getting to Mars for Real

We were excited. Now all we needed to do was build a space-qualified version of our robot, and it was going to go to the Moon. That is when the real world of politics entered the picture.

There was quite a debate over whether BMDO should really be putting rovers on the surface of the Moon. The Clementine mission to orbit the Moon had been justified on the grounds that it would be a good test of the navigation, autonomy, sensors, and reliability of the BMDO systems that were supposed to operate by themselves in space. It was harder to justify actually landing on the Moon. That surely was the preserve of NASA. Never mind that it was almost twenty years since NASA had last landed a payload on the surface of any body other than Earth itself.

I was not privy to the political maneuvers that went on behind the scenes. NASA had already announced that they had a lander due to be launched to Mars in late 1996. As of October 1993, when our Grendel had flown at Edwards Air Force Base, there was no rover associated with that payload.

About three months after our successful test BMDO announced that they would not be flying us to the Moon. At the same time, NASA announced that they would be sending a rover to Mars in less than three years! They had a little project going on out at JPL developing a new kind of low-cost rover. What would have been *Rocky 6* in their numbering scheme was now going to Mars. And it did, once it got renamed as *Sojourner* and flew as a "Faster, Cheaper, Better" package, to quote the new NASA jargon. *Rocky 6/Sojourner* was a rocker-bogey mechanism robot that was the current descendant of the small rover project that Colin Angle and I had seeded at JPL back in 1989.

On July 4, 1997, I was at JPL for the landing. The improbable landing mechanism of bouncing on inflated balloons worked, to the incredulity of many. *Sojourner* powered up successfully and traveled down its ramp and out onto the surface of Mars. Initially it was controlled very directly and closely, with the day's motion commands uplinked from engineers on Earth. After the primary mission of seven Mars days, and the secondary mission of twenty-one further Mars days, it was finally allowed to operate autonomously, under the behavior-based control approach that we had developed for our artificial creatures at MIT. Unlike every previous robotic space probe, *Sojourner* now had an agenda of its own and was operating autonomously without human control.

The first mobile ambassador from Earth to another planet, a creature constructed of silicon and steel,[13] was busy exploring. For the first time,

13. Actually much of it was aluminum rather than steel.

mankind had sent a robotic creature, an autonomous mechanical being, rather than a biological creature, as its vanguard representative.

Further Reading

Brooks, R. A. 1999. *Cambrian Intelligence*. Cambridge, Mass.: MIT Press.

Brooks, R. A., and A. M. Flynn. 1989. "Fast, Cheap, and Out of Control: A Robot Invasion of the Solar System." *Journal of the British Interplanetary Society* 42 (10): 478–85.

Kelly, K. 1994. *Out of Control: The Rise of Neo-Biological Civilization*. Reading, Mass.: Addison-Wesley.

Marr, D. 1982. *Vision*. San Francisco: W. H. Freeman.

Matijevic, J. 1998. "Autonomous Navigation and the Sojourner Microrover." *Science* 280: 454–55.

Pentland, A. P., ed. 1986. *From Pixels to Predicates: Recent Advances in Computational and Robotic Vision*. Norwood, N.J.: Ablex Publishing Corp.

4.   It's 2001 Already

On January 12, 1992, I held a party at my house for all my graduate students. We had champagne and cake and watched a movie. It was *2001: A Space Odyssey*. Most of my students found it slow and boring with not much in the way of special effects. But for me it was pure nostalgia. That movie, more than any other single event, had changed my life. In particular, the central character, the computer called HAL 9000, had inspired me as a teenager in Adelaide, South Australia, to dedicate my life to building intelligent machines. I still cannot watch it without my heart quickening and tears coming to my eyes often. It is awesomely inspiring to me.

In the movie it is mentioned that HAL 9000 was first switched on January 12, 1992—the book version says 1997. But 1992 was good enough for me, and I wanted to have a party to celebrate the most important imaginary event in my life—the day that gave life to HAL 9000 on the Urbana campus of the University of Illinois, under the guidance of Dr. Chandra.[1] HAL turns out to be a murdering psychopath, but for me there was little to regret in that. Much more importantly HAL was an artificial intelligence that could interact with people as one of them, using the same modalities that people use to interact with each other. HAL was a being. HAL was alive.

In the movie, HAL was a disembodied entity with superhuman intelligence levels. HAL was built into the spaceship *Discovery* and talked to the crew in English. HAL had lots of display screens all around the ship, but he got most input via speech and his glowing red cameras that were scattered throughout the ship. Speech and vision as input; speech, screens, and control of the ship as output.

As we ate and drank and chatted, it occurred to us that reality had failed to live up to the fiction developed back in 1968. Apart from the Macintosh interface, most people used a line-oriented DOS or UNIX interface to talk to their computers. Speech understanding and synthesis were both crude. Computer vision did not operate in anything like real time and worked only in very controlled circumstances. We had graphics on most computers but only black and white, no color. But no computer was able to carry on a conversation with people. No computer learned and developed from interacting with human teachers, as HAL had, learning to sing songs like a child. In reality, that day there was no chance that anyone was switching on a HAL-like computer that would grow and develop into an artificial being.

Unlike just about every other work of science fiction about computers, *2001* had not been overtaken by reality. It was still visionary. It was still way out. It was still completely imaginative.

One of my students at the 1992 party was Cynthia Breazeal, then working on getting Attila and Hannibal to walk like insects. On May 9, 2000, Cynthia delivered on the promise of HAL. She defended her MIT Ph.D. thesis about a robot named Kismet, which uses vision and speech

1. Dr. Chandra was a minor figure in *2001* but had a much more significant role in *2010*.

as its main input, carries on conversations with people, and is modeled on a developing infant. Though not quite the centered, reliable personality that was portrayed by HAL, Kismet is the world's first robot that is truly sociable, that can interact with people on an equal basis, and which people accept as a humanoid creature. They make eye contact with it; it makes eye contact with them. People read its mood from the intonation in its speech, and Kismet reads the mood of people from the intonation in their speech. Kismet and any person who comes close to it fall into a natural social interaction. People talk to it, gesture to it, and act socially with it. Kismet talks to people, gestures to them, and acts socially with them. People, at least for a while, treat Kismet as another being. Kismet is alive. Or may as well be. People treat it that way.

Reflections on HAL

A few months after my HAL party I went on sabbatical for the 1992–93 academic year. I wanted to understand artificial evolution and artificial life better. I had spent the last few years building insectlike robots. While they were much better robots than just about anyone else was building, they were not "humanlike," as was HAL. I had been thinking that after completing the current round of insect robots, perhaps we should move on to reptilelike robots. Then perhaps a small mammal, then a larger mammal, and finally a primate. But we had already spent ten years on insect robots. At that rate, given that time was flowing on for me, just as for everyone else, it was starting to look like, if I was really lucky, I might be remembered as the guy who built the best artificial cat. Somehow that legacy just did not quite fit my self-image.

As I pondered this and thought about HAL, I decided to try to build the first serious attempt at a robot with human-level capabilities, the first serious attempt at a HAL-class being.

I wanted to move away from mobile, complete creaturelike robots toward one that could interact with everyday objects. My first idea was a robot arm, with a camera mounted in the ceiling above, that would interact with objects placed in front of the arm. The ceiling camera would interpret the objects in the visual scene, and the arm would manipulate them, perhaps holding them up to the camera to get a better view.

I realized that I was falling back into the same mistake that artificial intelligence researchers had made again and again. The most important aspect of my insect artificial creatures had been their bodies in the world. A robot did not need an ability to avoid obstacles, but rather, with sensing of its visible environment, could much more easily be programmed to seek paths through empty space. The physics of its very embedding in the world provided a rich dynamics of interaction. In order to make it act intelligently, all we needed to do was nudge the dynamics in the right direction—we did not need to compute ahead of time every motion of every appendage and actuator.

Having a body provided a natural grounding. All the computations that might be done on the robot were in service of the robot in the world. Researchers who worked on abstract reasoning got as their result abstract reasoning. There was no way to tie it in to a physical robot after the fact. I have often tried to explain this mismatch and have come to the conclusion that work in abstract reasoning has as its only interface to the world conference papers written by researchers. Researchers are operating in an underconstrained environment, and as they follow up interesting research ideas, they are tempted, and succumb, to make their abstract world more interesting for their research ideas, rather than being faithful to the reality of the physical world.

My ceiling camera and robot arm needed to be part of an embodied robot, not just disembodied components mounted on a bench and a convenient ceiling. I needed to build a full artificial being, a robot with arms and a head that was a whole within the world.

But what exact form should this robot take? The almost automatic answer was that if the robot was going to have humanlike abilities, then it should have human form. But that, surely, was too simplistic. Was that really a valid reason?

There turned out to be two good arguments for building the robot with human form. The first is an argument about how our human form gives us the experiences on which our representations of the world are based, and the second argument is about the ways in which people will interact with a robot with human form.

We humans are not just products of our genes. We are also products of our social upbringing and our interactions with the world of objects. Our culture too is a product of our embodiment within the world.

The philosophers George Lakoff and Mark Johnson have argued that all of our higher-level representations of language and thought are based on metaphors for our bodily interactions with the world. Their arguments fill a number of books, so at best I can cite a few examples to give the flavor of the sorts of metaphors of which they speak.

Lakoff and Johnson start with primary metaphors that they claim are developed during childhood from physical and social experiences. For instance, they argue that affection uses warmth as a metaphor because a child is exposed to the warmth of a parent's body when shown affection. So our language becomes "they greeted me *warmly*." Likewise, importance has bigness as its metaphor, as in "tomorrow is a *big* day" because parents were important, were big, and indeed dominated visual experience. Difficulties are burdens, as in "I am *weighed down* by my responsibilities," because of the discomfort and difficulty of carrying heavy objects as a child.

Higher-level concepts are built as metaphors that are less direct than the primary metaphors but nevertheless rely on bodily experience in the world. For instance, for time we use a metaphor of moving forward, walking or running in a straight line. Thus the *future* is *ahead* of us, the *present* is *where we are,* and the *past* is *behind* us. If we were purely computational creatures without bodies and unable to move in any way, Lakoff and Johnson would argue that we would not have developed such a metaphor for time. It is not our only metaphor for time, of course. Sometimes we think of it as a flowing fluid, streaming past us, so that the *future* is *moving toward* us, what we are encountering *now* is *moving by* us, and the *past* has *moved past* us. Each of these and other time metaphors is rooted in an understanding of the physics of the world and how we can move about in it. That is abstracted then to form a metaphor with which we are able to manipulate more difficult concepts. Our language reflects these metaphors.

If we take this seriously, then for anything to develop the same sorts of conceptual understanding of the world as we do, it will have to develop the same sorts of metaphors, rooted in a body, that we humans do. For this reason it is worth exploring the building of a robot with human form and seeing just what sorts of metaphors we can get it to derive from its bodily experiences in the world.

Now, perhaps, in making this first argument I am guilty of cargo cult science. Cargo cults grew up among the Melanesian native populations of the Papua New Guinea islands during World War II. Having observed silver birds descending from the sky and unloading cargo, some native populations built facsimiles of landing strips with wooden radio towers and waited patiently for their own silver birds to alight. More dangerously some, such as the Biak Islanders in July 1942, carved wooden guns and then attacked massed Japanese troops. The guns of the Japanese were not made of wood, however, and the similarly shaped wooden totems of the Biaks were no match in the massacre that followed.

The robots we build are not people. There is a real danger that they bear only a superficial relationship to real people in the way that the wooden guns of the Biak Islanders bore only a superficial, and nonfunctional, relationship to the metal guns of the Japanese soldiers. Perhaps we have left out just too many details of the form of the robot for it to have similar experiences to people. Or perhaps it will turn out to be truly important that robots are made of flesh and bone, not silicon and metal. We will return to these themes in chapters 8 and 10.

The second argument for building robots with human form is that people will naturally know how to interact with such robots. We have been programmed through evolution to interact with each other in ways that are independent of our particular local culture. We make eye contact and we avert eye contact to signal whose turn it is to speak. We give encouragement that we understand as someone else talks, through nods and sublinguistic murmurs. We know when we are too close to someone from the way that person draws back. We know when someone wishes to engage us in conversation from the way he cocks his head toward us, waiting to get a signal via eye contact. All these things come naturally to us as we interact socially with people we have never met before. All these social cues were missing from HAL's red passive lenses. All these things would be missing from a robot camera attached to the ceiling. It seemed to me that building a robot with human form would allow us to explore these issues of subconscious communication.

By June 1993 we had started to build a *humanoid* robot. Cynthia Breazeal came up with the name Cog for our first robot: cog, an abbreviation for "cognitive," and cog, a tooth on the rim of a gear. "Cog" was a fitting pun, crossing over between the mechanical nature of our humanlike robot and the intellectual models we intended to implement upon it.

The Humanoid Explosion

When we started the Cog project, there were hardly any robots with humanoid form outside of science fiction. The one exception was at Waseda University in Japan. But now the situation has changed—there are humanoid robots in labs all over the world, and many of them owe their beginnings to Waseda University in Tokyo.

While artificial intelligence research in the United States and Europe tends to be done in computer science departments, in Japan it is usually done in mechanical and electrical engineering departments, in conjunction with the building of physical robots. Waseda is no exception to this.

In the early 1970s, Professor Hirokazu Kato of Waseda University started to build a humanoid robot. This work has blossomed into the Humanoid Robotics Institute, which has about a hundred researchers. The first humanoid robot they built was Wabot-1 in 1973, a robot that could walk a few steps on two legs, grasp simple objects with its two hands, and carry out some very primitive speech interaction with people. Wabot-1, however, was not an artificial creature. It did not exist in an environment and react to it. Rather, it was a demonstration machine, which when configured correctly would carry out one of its tasks without much regard to what was happening around it in the world. Wabot-1 was not a situated creature.

Professor Kato's next robot, Wabot-2, came to life in 1984. This was a situated robot, but in a very limited domain. Like Wabot-1, it had two legs and two arms, but Wabot-2 was restricted in very special ways. It could not stand, but sat, on a piano bench. Its feet were used to press the pedals of an organ, and its arms and hands were restricted to playing the organ keyboard. It had five fingers on each hand and could move its arms from side to side playing the keys. What made Wabot-2 more than just the mechanism of a player piano is that it sight-read music. Its head was a large TV camera, and when sheet music was placed on the music stand above the keyboard, Wabot-2 would read that music and play the piece. Although limited, Wabot-2 was a situated creature. Within its domain it existed and interacted.

Not much happened in the world of humanoid robots for a few years after Wabot-2. The Waseda team continued to build robots with humanoid form but mostly were concentrating on getting walking to work. There was not a premium put on making the robots situated, able to operate in the real world. Professor Atsuo Takanishi made great progress in understanding the dynamics of walking on two legs, but the robots he could afford to build were rather large and clunky, and were never able to work elegantly. Budgetary constraints would not allow Takanishi to build robots with sufficiently powerful motors and sufficiently light legs.

It turned out that there was something going on in humanoid robotics. Honda, the car company, had a ten-year project to build a humanoid robot, which they kept totally secret from Japanese academics, and everybody else, until they announced it in 1997. I had once been invited to give a talk at their Waco research center, just outside of Tokyo, in the early 1990s. Normally when I give such talks, people show me around and at least show me their already published work. On this occasion I was not let in past the security desk but instead was ushered to an empty room and asked to give my talk on six-legged walking robots to an anonymous group of engineers. I asked them what they worked on, whether it was robots or catalytic converters, but they told me they could not tell me what they did at this lab. I objected to being treated so uncollegially and tried to bargain some information out of them before giving my talk and leaving. The best I was able to get was to go round the room and have each researcher tell me what academic discipline his undergraduate major had been.

When Honda unveiled their P2 humanoid in 1997, quickly followed by their P3 in 1998, it was clear that they had put a lot of resources into their research program—over a hundred man-years of engineering had been devoted to the project already. The robots looked like persons in space suits. But in videos, at least, they walked in a very humanlike way. The Honda engineers had concentrated on walking—their first robot, the P1, had been just a pair of legs with hips and a heavy weight on top. The humanoid robots also had an upper torso, a battery back-pack, two arms, and a head. Only the legs were operated autonomously, and they followed the ZMP walking algorithm developed by Takanishi at Waseda. The arms and head were operated by a person wearing a virtual reality suit. The robot did walk extremely well, including the ability to climb stairs. It could not navigate itself and avoid obstacles, however—that is all done by a remote operator with joystick control. Honda's robots are not really artificial creatures in the same way that the early Waseda robots were not. They are complex electromechanical devices that can play out a little piece of behavior that is not really in response to the situation in which they find themselves in the world—the problem is that they do not actually find themselves in the world, as they are controlled by a person rather than being autonomous in any sense.

In early 2001, Honda announced yet another, smaller version of the humanoid, known as Asimo.[2] It is being marketed as a remote control device for amusement parks. It is a small humanoid, the size of a ten-year-old child. Under remote control by a person, it can walk, reach, and grasp, and via a closed-circuit television system the operator can talk to people and see their expressions. It does not operate as an autonomous robot, but it is intended that the audience should think that it is.

In the last few years there has been an explosion of work on humanoid robots. The Waseda team brought out a host of new humanoids. Tokyo University and the Electrotechnical Laboratories in Tsukuba, and the Advanced Telecommunications Research Lab near Kyoto, all had major projects. Sony, following their successful dog robot AIBO[3] demonstrated a 50-centimeter-tall humanoid, also walking under the Takanishi's ZMP algorithm. Fujitsu Research Labs and Tokyo University both constructed similar walking robots. At the MIT AI Lab, Gill Pratt and his students built new bipeds, one a dinosaur, but the other very much a humanoid set of legs, called M2, but with a much more powerful set of walking algorithms than the other groups. Many groups in England, Germany, and the United States built humanoid bodies, arms, and heads. Most of these groups have tried to move beyond the Honda model of expensive puppet, doing as commanded. All tried to build

2. Surely this is tribute to the science fiction writer Isaac Asimov.
3. For more on AIBO, see chapter 5.

sensory systems that let the robots be aware of their surroundings and to act in some appropriate way. Many of these humanoid robots are able to sense when people are present and to change modes to interact with them appropriately. Some of the walking robots are able to avoid some obstacles by themselves. Pratt's robots are able to walk over rough and unmodeled terrain. Some of the robots have sophisticated eye movements that let them gather perceptions from all around them. A few of the robots are able to reach out and grasp objects that they perceive with their computer vision systems.

None of these robots has the finesse or intelligence of HAL at this time. But they have shown that building humanoid robots is not a completely crazy proposition. The technology that we have today can let us get partway toward the ultimate goal. And we are not limited by either computation power or intellectual ideas. We are limited only by time, and as we spend more time on our creations, we make them better and better.

Asimov's Laws and Reality

During the 1950s Isaac Asimov wrote a series of books about humanoid robots, starting with *I, Robot*. The main characters were robots, some of which were almost indistinguishable from people. The robots were manufactured to obey three laws, which have come to be known as Asimov's laws.

1. A robot may not injure a human being, or, through inaction, allow a human being to come to harm.
2. A robot must obey the orders given it by human beings except where such orders would conflict with the First Law.
3. A robot must protect its own existence as long as such protection does not conflict with the First or Second Law.

The plot device in many of the books and short stories that Asimov wrote about these robots centered on situations in which there were logical inconsistencies in the application of the laws. This always led to some unexpected behavior on the part of the robots as they struggled to obey the three laws.

Journalists, and others, have often picked up on these laws. In fact, the robot played by Robin Williams in the 1999 movie *The Bicentennial Man* projected these laws holographically when it was first delivered to its new owners. Journalists often ask whether the robots being built today are built to obey these three laws.

The simple answer is that they are not. And the reason is not that they are built to be malicious, but rather that we do not know how to build robots that are perceptive enough and smart enough to obey these three laws. At first sight, the laws seem innocuous and plain common sense. However, upon closer examination they turn out to be very subtle. Of course, Asimov knew this, as he played on that subtlety for his plots. Perhaps, though, he did not realize just what a perceptual load these laws put on a robot.

Up until 1998, or perhaps even 1999, no robot had a chance of obeying the first of Asimov's laws, because no robot had any way of even detecting a human. Except for Polly, all the robots mentioned in chapter 3 at best treat humans as obstacles and try to avoid them. Even Polly avoids them except in the special circumstance of their appearing as a vertical static object in a corridor.

Why is perception so hard? It seems the most effortless of things to us. We open our eyes and we are aware of what is in the world. We see people, objects, spaces, indoors, outdoors, our own hands. The vast majority of people can do this. Most other mammals seem to be pretty good at vision—our pet dogs and cats certainly act as if they see many things in the world. Birds too seem to operate very much on vision. These hypotheses have been well confirmed by careful experiments. Many other animals, such as wasps, bees, and many reptiles, also have acute vision. Although their vision has improved drastically over the last two to three years, today's robots are still extraordinarily bad at vision compared to animals and humans.

The fact that computer vision is difficult came as a big surprise to artificial intelligence researchers. In 1963, Larry Roberts[4] completed the

---

4. Later in the sixties Larry Roberts went on to be one of the pioneers of the ARPAnet, which ultimately turned into today's Internet. He is still active in the field of networking.

world's first doctoral thesis on computer vision at MIT. Using an early MIT computer called the TX-0, Roberts was able to distinguish simple white polyhedral objects on a black background. This early success was followed by little progress for a few years. Marvin Minsky, cofounder of the MIT Artificial Intelligence Laboratory and a consultant to Stanley Kubrick on 2001 back in 1967 and 1968, helping him to envision HAL, decided to get the pesky vision problem solved once and for all in 1966. Although everyone knew that playing a world-class game of chess was a long way off, perhaps ten years, surely vision could not be that hard. Even stupid people could do it well. Marvin and Seymour Papert assigned an undergraduate named Gerry Sussman to work on it for the summer.

We still have not gotten very far with that summer project here in 2001. We have computers that can beat the world chess champion, but we do not have computers that can see very well at all. Marvin completely missed the mark, and this had a big influence on how he saw HAL would be in the year 2001. Unfortunately, Marvin still did not get it after his failure to solve vision in a single summer, and he continued to belittle what I consider to be the truly hard problems for the next thirty years. Those hard, unsolved problems include such simple visual tasks as being able to reliably recognize a cup as a cup, a pen as a pen, and an automobile as an automobile. The problems include being able to patch together a stable worldview when walking through a room. And the problems involve being able to understand the sounds of a slamming door, a singing bird, or the heavily accented words of a recent immigrant. And the hard, unsolved problems include being able to see an object, pick it up, and roll it around in a hand. I once heard Marvin in the mid-eighties complaining at an artificial intelligence conference that too many people were working on vision and robotics, and it was as though those misguided people were working on daisy wheel printers. Daisy wheel printers were a particular technology for printing on paper just before laser printers became common, and Marvin's point was that computer vision and robotics was just input-output. The real problems in artificial intelligence were things like thinking. After all, even stupid people could do the input-output part—hearing, seeing things, and manipulating them.

There seems to be a straightforward explanation of how people see.

They open their eyes, receive and input like a video camera does, or like scanning a photograph, and then they see what is out there in the world. If we shut our eyes, we can recall what we saw, and even answer questions about things that we had not been "conscious" of seeing—this after all is the basis for eyewitness accounts in legal trials.

It turns out that this understanding of how we see is almost completely wrong. Our human vision systems are at the same time much worse and much better than this. Furthermore, the mechanisms that drive our eye movements so that we can perceive the world are also used for many crucial aspects of our basic social interactions. Seeing cannot be understood without understanding human social interactions, and likewise social interactions cannot be understood without understanding how we see. As in so many other ways, we humans are products of our embodiment in the world.

The Basis of Seeing and Behavior

Seeing is not a passive process. Rather it is an active ongoing activity. Our eyes move rapidly from place to place, getting only scraps of information about what is out in the world. Our brain fills in lots of details that are perhaps not actually seen in a process that is not yet fully understood. The result is that we feel as though we are seeing a stable view of a static world in full panoramic detail, requiring only our conscious attention to bring out individual details.

Our eyes are not much like digital cameras. Digital cameras have a uniform resolution across their 35-or-so-degree field of view. Modern video digital cameras may have about a quarter of a million picture elements, or pixels, each of which can see equally well in red, green, and blue. It turns out that this nice technical fact about cameras has been bedeviled by some compromises made for TV transmission. Although the cameras themselves are usually good at color discrimination at every pixel, the standard way to transmit video images throws away a lot of that color information. There are two reasons for this. First, the original televisions were only for black and white images. When color television was introduced, it was necessary to retain backward compatibility with existing receivers and transmitters—people with old televisions had to be able to see programs broadcast in color as though they had been broadcast in black and white, and people with new color television sets had to be able to view programs transmitted from old television stations that had not yet switched over to color signals. The color part of the signal had to be hidden in some unused space of the original black and white protocol, and there was not much room to fit it. Second, humans are not so very good at color perception, and so people did not miss having a high-bandwidth color signal displayed on their televisions— everything looked just fine to them with the reduced color that could fit in the signal. Although this was a clever technical solution at the time, it still plagues computer vision researchers because so many cameras collect good color signals, then throw them away before they output them in standard transmission format.

The human eye, by contrast with a video camera, has a field of view spread over about 160 degrees horizontally (a little more vertically). The retina has about 100 million brightness receptors, called rods, and about 5 million color receptors, called cones. These are not at all uniformly spread out, however. There is a very distinct region in the center of the field of view of each eye called the *fovea*. This 5-degree-wide region has a much higher density of receptors than the rest of the eye and is much more color sensitive. In fact, in the central part of the fovea there are no rods at all, just cones, and there are no blood capillaries overlaying the receptors in the light path as there are in the rest of the retina. The outer edges of the field of view are called the *periphery*.

It is easy to do some experiments with your own eyes. Shut one eye and stretch out your two arms in front of you, wiggling your index fingers. Look straight ahead and swing your arms outward until each hand gets right to the edge of your field of view—just at the point where you lose sight of your waggling index fingers. Your arms should now be making an angle of about 160 degrees with each other.

The reason you had to wiggle your fingers is that you cannot see very much at all besides motion out in your extreme periphery. You have almost no color vision there and cannot recognize objects. There are too few receptors for that, and the details are lumped together as signals sent from your eyes to your brain. There are about 1.5 million nerve fibers that run from the retina back to the brain to a region called V1. For the foveal area there are about 3 fibers per receptor, but in the periphery the ratio is 1 fiber per 125 receptors. Half of V1 is devoted to the central 10 degrees vertically and horizontally of your visual field—half of V1 is thus devoted to just 2 percent of the visual field, the central field including the fovea.

Before we leave the retina, it is worth pointing out that it is incredibly badly designed. Not only does the blood supply run over the surface of the light-sensitive elements, but the nerves that transmit the information to the brain are also in the path between the lens and the receptors. In the much better designed octopus eye the cable of nerve cells from the brain is attached to the back of the retina and picks up the image information by spreading out over the back of the light receptors. In mammals the cable comes from the rear and punches a hole in the retina only fifteen degrees from the center of the fovea, then spreads out over the surface of the retina. This leads to a blind spot we all have in each of our eyes. When we look in any particular direction with one eye we are completely blind in one spot.

Normally we are completely unaware of our blind spots in each of our eyes, but a simple experiment will reveal it. Get a white piece of paper and draw a good solid X in one place and a circular spot the size of your little fingernail about 15 centimeters (or 6 inches) to the left of it. Now hold the piece of paper in your left hand out at arm's length. Close your right eye and with your left eye stare right at the cross you have drawn. You should be able to see the dot off to the left. Keep your eye fixed on the cross and slowly move the paper closer to you. At some point the dot will disappear, then reappear as the paper gets even closer. Right at the point it disappears is when it is in your blind spot. Turn the paper upside down and repeat with your left eye closed and your right eye fixated on the cross to find your other blind spot.

If we stop to think about this for a little bit, we see (look at our metaphorical language—oops, did it again with "look") how remarkable our normal interpretation of looking at the world is. Our internal self-reflection on how we see tells us that when we look at the world with our gaze fixed in one direction we see things right at the point of gaze in high resolution. The resolution of everything else in our field of view falls off toward the periphery. Our introspection is hard put to let us know that our color perception falls off fairly drastically at the same time. And most of us have gone through our whole lives never even noticing those blind spots. We are totally unaware of them. Imagine if our television screens all had a blank spot of similar size in a fixed location away from the center of the screen. We would notice it in a flash, and be aware of it whenever we watched a TV program. Our awareness and our sensory systems are not quite the same thing.

It is not too hard, once it has been pointed out to us, to notice that we do a lot of steering around of the fovea. The human visual system does this so that the high-resolution color-sensitive part of the eye is looking at something that is currently behaviorally relevant. In turn, evolution has taken advantage of our eye motions to form the basis for many of our social interactions. But before we investigate that a little more, we will linger on eye movements a while longer.

Our eyes are capable of rapid point-to-point motions three or four times per second. Glance up from the page and back down again. Each motion your eyes just made is known as a *saccade*. During a saccade your motion vision system shuts down so that the world does not appear to be rapidly slipping out of your control. And there is no visual feedback to your brain during a saccade. A layer of neurons in your *superior colliculus*, part of your brain, corresponds to a two-dimensional panorama in front of you. As a set of neurons in a particular region is excited, your eyes saccade to the corresponding place. If their target was some visual feature that had been detected previously, there might be a secondary small motion to correct after the large saccade. If you started wearing prism glasses that compressed everything into half the normal angles, your saccades would not be very accurate at all. After a few days, however, your superior colliculus would adjust and you would be able to saccade just as well as before. Until you took your glasses off again. Saccades take about 60 milliseconds. During that time we are completely blind. Notice how we do not notice that. Again, our awareness and our sensory systems are not quite the same thing. Notice also that we can have voluntary control over where we saccade: you can choose to look up, then left, then right, then back to the page. But most of the time it happens without any awareness at all on our part. In fact, most people have never even thought explicitly about how their eyes saccade—but it turns out that our saccade behavior is vital to our ability to socially interact face-to-face.

Now it is time for another experiment. Lift your eyes from the page and pan them slowly and smoothly from left to right. You cannot do it. Instead, your eyes went through a series of jerks, saccades, from place to place. But next time someone walks by, or if you can see cars passing on the street, you will be able to smoothly track them. Your eyes will scan from left to right completely smoothly without any of those jerks. This is known as *smooth pursuit*. Here now is something that we can only do involuntarily! You can choose what to track and trick your eyes into moving smoothly, locked on to the image of something moving in the world, but you cannot make your eyes move smoothly themselves.

Our eyes have more tricks in the way they move about. There is the *vestibular-ocular reflex*. When your head moves, your inner ears measure the amount of motion and compensate for it by moving your eyes so that they stay pointing in a constant direction. This, again, works without visual feedback, so even if your whole body is suddenly and unexpectedly rotated a few degrees when you have your eyes open in the dark, your eyes will compensate and stare unseeingly at the same invisible place. And the trick with the prism glasses works here too—after about a week of wearing them your eyes will compensate in the dark as though they were seeing something through those Daliesque distorters. The vestibular-ocular reflex is not present when you are watching a movie on a big screen and the camera has undergone rapid motion. When you sit in the front of a bus and it speeds around a tight corner, the world seems stable as the reflex compensates for your motion. When you watch a movie where there is a camera sitting in the front of the bus, it feels like the whole world is careening out of control, because your brain is not getting fed the vestibular signal about the motion of the bus.

Your eyes also verge on objects. When you look at something in your hand, your left and right eyes are not looking in parallel directions. Rather, each of them is looking directly at the pen, for instance, that you are holding in your hand. For an adult that means that they are roughly 6 degrees off parallel and toward each other. The pen is at the center of the image in each eye, and by comparing the image around the center it is easy for our brains to work out what parts of the local scene are slightly in front of—if the images are closer together there—and slightly behind—if the images are further apart. This is known as stereo vision, and about 90 percent of adults are able to do it effortlessly and directly. Some people never develop this ability naturally, but they are still able to reconstruct depth by the relative way things move in the verged images on the two eyes as their head moves slightly in the world. As we saccade around our visual world, our eyes verge and our brains extract rough depth information by measuring the angles between our eyes. It turns out that we also visually measure the angles between the pupils on other people's eyes when we look them in the face, and we are able to roughly estimate both where they are looking and how far away they are looking. Note that these two skills—determining depth by verging our own eyes, and estimating the depth at which another person is focusing his attention—are completely independent, but clearly evolution has taken advantage of our own consistencies to teach us how to do this.

Each of these aspects of the human vision system are very machine-like in the way they operate. Usually they operate without our conscious control, and while we can exert conscious control on some of them (e.g., where we saccade) we cannot on others. During the last few years many research groups have managed to replicate these mechanical aspects of human vision in robots. Each of the subsystems is fairly easy to understand, and when we put them all together in a robot, they work pretty much as they do when combined in a person.

It is much more difficult at this stage of our technological development to duplicate the next stages in the human visual system. In contemplating all the mechanistic things that our human vision system does, it is worth stepping back into our subjective selves. The world around us is a stable place. We walk around, we sit down, we glance around. The world does not move. It stays in one place. We know about the painting on the wall behind us and have no trouble glancing over our shoulder at it. We know just where it is. Our eyes tell us exactly where everything is, and how it sits in the world. But at some level this is an illusion. Our visual systems are continuously jumping around all over the place, steering the sensitive parts of our eyes to look at things in a serial fashion. We have big holes in our eyes but do not seem to notice. Out of all this our brain constructs a nice, stable view of the world, and it seems to be in three dimensions. It seems as though, as our eyes flick around, we are building up a nice model of the world inside our heads, and this is what our brain has conscious access to. That is why we can all be good witnesses in a court of law, faithfully telling the jury just what we saw that fateful day. It is all sitting right there in our memory, a solid reconstruction of the external world as it was.

It turns out that this is not quite right either. We know this from the experiments of Alfred Yarbus, a Russian psychologist active in the fifties and sixties. He developed the first apparatus that could track exactly where people were centering their eyes as they looked at a photograph. When Yarbus gave someone a photograph of a person's face, the subject would invariably scan back and forth between the two eyes, down the nose, back and forth across the lips, and then scan around the boundary of the face. People tend not to foveate on the cheekbones, despite our understanding of their importance to beauty. The stunning experiments that Yarbus reported, however, were ones where he gave the same photograph to many different subjects, having asked them in advance certain questions about the photograph they were about to see. Different people had qualitatively similar patterns of eye motion on the same question, but each question reliably elicited very different patterns from other questions.

In the best-known example the subjects were presented with a copy of the Russian artist Ilya Repin's 1884 painting *Unexpected Return*. There were two women and two children in a room, with someone who was clearly a servant outside the door letting in someone who was an unexpected visitor. If subjects were asked to estimate the age of the people, they would look at each person in turn, intently scanning their gaze around each person's face with very little looking elsewhere. If asked to remember the clothes worn, they would again scan each person in turn, and then scan up and down the people's bodies. When asked to estimate how long the visitor had been away, they would go back and forth between the people's faces, just taking a single fixation on each face, then on to the next face, returning for one quick look at each face again and again. All these patterns were very different from each other and very different from the general scanning around that people did when presented with the picture but without explicit questions.

It seems that rather than "photographing" the picture in their mind, as proponents of photographic memories would claim, people actively search for and store information relevant to some task. Dana Ballard and

Mary Hayhoe and their students at the University of Rochester decided to investigate this hypothesis further.

Ballard and Hayhoe set up an experiment in which people were to make copies of two-dimensional layouts of Lego blocks. They were given a pattern of colored blocks to look at on their left, a construction site on their right where they were to make a copy of that pattern, accurate in both shape and colors, and a pile of blocks between and closer to them. They were instructed to pick up one block at a time from the pile and place it in the copy they were building.

Now suppose that the photographic memory hypothesis was correct, the hypothesis that most of artificial intelligence has worked with for over forty years. In that case, one would expect the subjects to look at the pattern and store it in their minds. Then they would look at the pile, pick up a block, saccade up to the copy region, place it down, then saccade back to the pile for the next pickup, repeating until the task was done. That is not at all what happened. The subjects would look at the pattern, look at the pile, pick up a block, look back at the pattern again, look at the copy region, place the block, look back at the pattern, and then back to the pile again. This is quite a surprise. Even if one thought that the whole of the pattern was not getting stored in the subject's head, one might think that they would look at *pattern, pile, copy, pattern, pile, copy*, etc. But instead it is *pattern, pile, pattern, copy, pattern, pile, pattern, copy*, etc. It seemed that perhaps people could not remember at the same time both the color of the block that they needed to pick up and the position where they were supposed to place it. Instead, their saccade strategy seemed to indicate that they were only remembering one thing for one block at a time.

The next step for Ballard and Hayhoe was the beautiful step in experiment design and execution. If it were really true that people were only keeping one piece of information in their minds at a time, would it be possible to change other things in the world and not have them notice? Scientists already knew that it was possible to fool invertebrates this way. Certain digger wasps dig a tunnel as a nest, then go off and anesthetize a caterpillar that will become a living food source for its young. It drags the caterpillar back toward the nest, then when it is about 5 centimeters away, it leaves it there and goes down the nest, to check whether anyone else has taken up residence there. This makes good sense, because it may have been off hunting for the caterpillar for quite a while. But it appears that the wasp is not reasoning this through at all. Rather, evolution has programmed the wasp to always stop about 5 centimeters from the nest and check the chamber, no matter how long it has been away. The experiment to demonstrate this is rather simple. While the wasp is down the hole checking, simply drag the caterpillar a few centimeters farther away from the nest. The wasp returns and soon finds its bounty by smell. It drags it toward the nest, but as it crosses the magical 5-centimeter threshold, it stops and goes to check the nest again. This can be repeated until either the experimenter or wasp wears out. If the caterpillar is not moved, the wasp comes back up, grabs it, and drags it down the hole, where it proceeds to lay its eggs. Could humans doing a Lego construction task be similarly feeble-minded?

Ballard and Hayhoe built a new virtual version of their construction task. Now everything was done on a screen with a mouse and colored rectangles instead of actual Lego blocks. As before, they monitored exactly where the human subjects saccaded their eyes. But now they added a devilish twist. While the subjects were not looking in some particular direction, they changed the color of a block in the pattern, being careful to keep it consistent with the rest of the virtual world in which the people were working. The human subjects did not notice if they were on their way to look at the pattern before picking up a block, even though they had already looked at the pattern many times. If they already had picked up a block and were on their way back to look at the pattern again, and the block whose color was changed was the one they were in the middle of copying, they would pause only a little. Then they would choose some other pattern block consistent with the one they had picked up and copy its location instead. The subjects really were using the world rather than their memory to do these simple tasks. They would not notice when the world changed out from under them. Surely you would not be that feeble-minded, would you? It just does not seem possible.

It turns out, however, that this is just one of many similar sorts of things to which we can subject ourselves that seem totally against our experience of the world. And in some sense they are all against our experience of the world. The world just does not change like that. As a result, evolution has not had to wire us up to notice such changes. They usually do not happen, so there is no need to pay the cost of checking things that do not happen. Ron Rensink, Kevin O'Regan, and Jim Clark have a series of even simpler experiments that regularly surprise my students when I show them. They have made tapes of images in which something drastic changes in an image regularly. The "game" is to watch the tape and try to notice just what it is that changes. It takes the average person 24.0 seconds to notice that one of the four engines of a Boeing 747 is disappearing and then reappearing in a picture of the airplane. People sit and stare at it and just cannot see what is changing, or indeed that anything is changing. Only when they finally happen to be looking right at the engine when it vanishes or appears do they notice it—after that they cannot force themselves to ignore it. In a kitchen scene it takes 32.5 seconds on average to notice that the color of the dishwasher is changing from brown to white and back.

So much for reliable eyewitnesses at criminal trials.

The human visual system turns out to be a complex arrangement of partial solutions to difficult problems. It is the product of evolution, not of careful engineering design, and that shows through in just about every aspect of it. The system is much more than just a camera. It is an exquisitely balanced system of interacting sensory motor control loops and is very active in its behavior in the world. The goals and current behavior of any person have a large influence on how their vision system is operating at any particular moment. Our vision systems are not as good at seeing things as we introspectively may have thought. Somehow our brain patches over all these difficulties and lets us imagine we are living in a stable world where we see and understand almost everything.

HAL 9000 had none of the motor skills that human vision relies upon. HAL had fixed cameras with lenses looking out on the world. With sufficiently advanced technology one could imagine that those cameras were very large arrays looking at the world through fish-eye lenses. Perhaps they could have many thousands of pixels on a side rather then the few hundred in the common cameras of the actual year 2001. Then there would not be the need to swing the cameras around physically as we are forced to do with our eyeballs. Rather an internal focus of attention could shift around and HAL could concentrate on the appropriate parts of the image for the appropriate task at hand. But what HAL would still miss is the thing that it most obviously missed in the film. The ability to have natural social interactions with people, social interactions based on the evolutionary hijacking of the mechanism for collecting information about the world and reusing it as the mechanism for regulating social intercourse.

Visual Behavior and Social Interaction

When we look at another person, we get many cues about what is going on inside their heads by seeing how their eyes move, and what they are looking at.

Since vertebrates all share steerable eyes with a foveal region (as do many other species, including spiders), it has made sense for evolution to capitalize on understanding gaze direction. A fairly basic principle is that if something is looking at you it may be going to eat you. Thus many animals have gotten very good at understanding when they are being looked at by other animals, independently of whether they evolved with a particular species in their ecological niche.

Hog-nosed snakes, for instance, play dead when people are around, and play dead longer if people are looking directly at them. Plovers, which nest on the ground, fake having a broken wing to entice predators away from their nests and their young, and will carry on the charade longer if a human is looking directly at them rather than just being present in the vicinity. Neither of these behaviors provides any evidence of conscious behavior or deception on the part of the snake or bird. The interesting point is that their behavior changes depending on the gaze direction of humans present.

Between people, gaze direction and gaze-direction determination are crucial foundational components of how we interact with each other. We use another person's gaze direction to determine what they are paying attention to. If we are talking to someone and their eyes are scanning the newspaper in front of them, we realize that they are not paying attention to us, even if they are occasionally saying "Uh-huh" and nodding as we talk.

Human infants develop an understanding of gaze direction in a series of steps. Other animals follow the same series, but most of them do not advance as far as humans.

At a very early age an infant can tell when their parent is looking at them, and they stare back into their parent's eyes. A little later, when the parent breaks eye contact and looks somewhere else, the infant starts to search for something nearby of interest but is not able to determine whether the parent is looking to the right or to the left, just not looking directly into the eyes anymore. By about nine months of age the infant is able to follow the direction of gaze of their parent in looking for just what it is that the parent is paying attention to. By twelve months the child is able to estimate the distance at which the parent is looking by measuring the vergence angle of their eyes and can saccade directly to the object of interest. Still later the infant is able to determine if their parent is looking outside the infant's field of view and will turn their body to look over their shoulder at whatever it is of interest.

The movements of another person's eyes are critical for us to understand what it is they are paying attention to. We are extraordinarily good at understanding people's gaze direction and eye movements. Many of us have had the uncomfortable feeling that the person next to us on an airplane is reading our magazine or computer screen over our shoulder. We can hardly make ourselves conscious of the person's face or eyes while we keep our eyes locked ahead, but somehow our peripheral vision delivers a message to us about just where they are looking. Again, evolutionary pressure would have made us good at understanding this. Perhaps they are about to steal our food, or would have done so in evolutionary times—there does not seem much point to it with modern airplane food.

Gaze direction has been reused by humankind's social interaction system as a cue during many of the daily face-to-face interactions that we have. When we approach a bank teller (if we still do), they may have their head slightly down and eyes pointing down at the workplace in front of them, still shuffling paper or pressing keys from the last transaction. This tells us that they are not yet ready to interact with us. We wait, and eventually they look up and make eye contact. That is our signal that our transaction can begin. We routinely go through this ritual many times a day as we approach people in offices, stores, cafeterias, ticket booths, and coffee shops. We do not consciously think about it, it is just how humans operate across most cultures in the world.

When we are having a conversation with someone, we also use our eyes for turn-taking. Now we can converse with people without face-to-face contact, as we know from using telephones. There we detect a pause to know when it is our turn to pick up the conversation, but often we make mistakes and interrupt each other, especially when talking to people on the other side of the world, where there is a perceptually obvious delay due to the time it takes signals to travel the distance necessary. Face-to-face, however, we make fewer mistakes, and we use the making and breaking of eye contact to know when it is our turn to speak. If the other person is looking away as they speak, they may bring their eyes back to our face as they pause, letting us know that it is an opportunity for us to take up the dialogue. Or if the person is looking into our eyes as they speak, they may glance away and then back to let us know that it is our turn.

We also make use of gaze direction as we interact with people, to know what they are paying attention to. Often we will say something that makes no sense by itself, but in the context of what the other person is looking at, it makes perfect sense: "No, that one just to the right." We also maintain some understanding of what people know based on where they have been looking as we are interacting with them, and can predict that their state of knowledge of the world may not be accurate and that we need to help them out with some verbal information: "No, I just put it up on the shelf," or "I just threw it in the trash folder." Again, these are things about which we are not normally conscious. It has become part of the fabric of our social interactions, but it is often very much based on our visual abilities to perceive and track the visual attention of others.

One way to build a robot that can interact socially with people in a completely natural way is to build it with a vision system that works like that of people, and with eyes that can saccade and verge, and that look like human eyes. In that way, people will be able to understand how to interact with it. This was one of the insights I had had back in 1993, and which had been part of the decision process that led to building our first humanoid robot, Cog.

Building an adequate vision system for Cog was one of our first preoccupations. In 1993 it was still hard to get enough speed from off-the-shelf computers to do real-time computer vision, but by the end of the nineties that was no longer the case. We now usually have enough computer power to run whatever algorithms we choose. It is finding the right algorithms that is difficult.

We have been able to duplicate the mechanical aspects of the human visual system. Rather than have very wide-angle cameras for Cog's eyes, with smoothly varying resolution and a fovea in the center, as in the human eye, we use an approximation. Each of Cog's eyes is really two cameras. One has a very wide-angle lens so that Cog can have a peripheral view, and one has a very narrow-angle lens to give Cog a fovea. Unlike a human, it does not have a blind spot, but it does have a new problem. In each eye it must compensate for whatever misalignment there may be between its fovea and its periphery.

Each of Cog's two camera eyes are mounted on gimbals that can pan and tilt so that Cog can look around. Its head and neck gives it even more scope for its visual exploration of the world. Just as in people, when Cog looks off in some direction, its head soon turns in that direction so that its eyes are roughly centered in their view direction relative to the front of its head.

Cog is able to saccade from one place to another. It is constantly updating a motor map in its own superior colliculus, so that it accurately saccades to what it chooses, in spite of wear in its motors, or mechanical readjustments made by a graduate student. It is able to smoothly follow somebody walking in front of it. Because its head might be at any angle relative to the person's trajectory, this is not just a matter of panning its eyes from side to side. Instead, the independent pan-and-tilt motors need to coordinate to form the appropriate path. Cog's head also has a gyroscope to play the role of the inner ear detecting its own head motion. Cog's version of the vestibular-ocular reflex allows its eye motions to successfully saccade, smoothly follow, or just fixate on a distant object, no matter how its head is moving independently.

The difficult thing is how to process the images. My earlier description of the human visual system left out those details because we do not really know how the human brain does it. We have certain understandings how the first parts of the processing work. We know about the color sensitivity of the cones, and how the retina is able to detect brightness edges—places where the image changes intensity, such as at a shadow boundary or at the vertical corner of a building revealing vegetation behind. We also know how the retina is sensitive to motion. But once the signals are sent back to the first visual areas of the brain, we really do not know what processing is being carried out.

People working in computer vision for the last forty years have had to try to come up with theories about how to make algorithms that can understand a pattern of intensities and turn it into rich descriptions of what it is in the world appearing in front of the camera. It has turned out to be much, much harder than anyone expects. What appears to be a sharp boundary to our eyes and vision system combined does not necessarily show up in an obvious way in the data that comes from a digital camera. If we see a pen lying on a desk, we can see a sharp boundary between the pen and the desk. But often when we look at the intensities of light from each little square pixel in a digital image, there is no clear boundary. Pixels corresponding to parts of the pen, and parts of the desk, just two or three pixels apart, may have exactly the same intensity values. Somehow our brain is getting a much more global understanding of what is going on, and it then *perceives* the boundary.

In the early days, when computers were slow, computer vision researchers tried to make very clever algorithms. Over the last few years, however, a lot of brute-force algorithms have become relatively successful. These brute-force algorithms do a lot of simple computations everywhere in the image, without very sophisticated underlying models of how the image might have been formed by the physics of the world. Algorithms that would have been looked down upon during the seventies and eighties as unsophisticated have become popular. The reason they are popular is that they work. Sometimes they do not do as well as a more sophisticated algorithm on an individual image, but when running algorithms on a vision head, such as Cog's, with actively moving cameras, there would always be another chance to process a slightly different set of images a thirtieth of a second later. Any set of images with particular errors would be replaced by another set of images before very long. By smoothing the results out over time, people were able to build very reliable vision systems that operated in real time on a constant flow of images.

These vision systems, however, were not, and are still not, all powerful. We still only know how to do certain subclasses of vision algorithms.

The things that computer vision is currently good at include:

1. Detecting human faces in a scene.
2. Recognizing human faces from a relatively small library of faces as long as there is a full frontal view.
3. Finding eyes on faces.
4. Tracking moving objects from a stationary camera.
5. Determining the rough three-dimensional structure of a scene over distances of two or three meters.
6. Registering a detailed geometric model, even of flexible variable structures like human organs, to three-dimensional data.
7. Picking out saturated colors, and skinlike colors (across all human races—the underlying pigmentation properties are the same and can easily be extracted).

The sorts of things that computer vision is still not good at include:

1. Compensating for camera motion in tracking objects.
2. Recognizing whether a face is that of a male or a female, or a person who is young or old, or just about any other kind of discrimination.
3. Determining the gaze direction of someone with high accuracy.
4. Recognizing people from nonfrontal views.
5. Recognizing people as they age, wear a hat, shave, or grow a beard.
6. Recognizing what people are wearing.
7. Determining the material properties of something that is viewed.
8. Discriminating a general object from the background.
9. Recognizing general objects.

The truth of the matter is that we have no computer vision system that is at all good at recognizing that something is a cup, or a comb, or a computer screen. Our computer vision systems can do a few things with great skill, but still after forty years of effort they are not good at the things that we humans and many animals do effortlessly.

Because of the increase in computer power over the last thirty years, we can no longer blame a deficiency there for our poor computer algorithms. It is clear that we must be missing something fundamental in the way that vision in humans is organized, although almost no one will admit that. Every researcher is convinced that what they are working on is the most important thing in computer vision, and all manner of riches will soon flow from it.

While these troubles in the land of computer vision persist, those of us building intelligent robots have to work around them. Our robots float in a strange perceptual space in comparison to the spaces that you and I inhabit. There is little constancy, and only motion and a special class of objects (e.g., faces) can be recognized at all. In imagining what it is like to be one of our robots, the best analogy is that of a strange, disembodied, hallucinatory experience.

Kismet

Cynthia Breazeal worked on Cog, designing it mechanically, and building its first set and second set of brains. She built its first superior colliculus, so that it could saccade to motions that it saw in its periphery; and she worked with another graduate student, Robert Irie, to extend the capability to sound, so that Cog could saccade its eyes to wherever it heard a sound.

One day we took some video of Cynthia interacting with Cog. It was running programs that she and a number of other graduate students had written. The point of the videotape was to show that it was safe to interact with Cog, up close. Cynthia held a whiteboard eraser and shook it. Cog would saccade to it and then reach for it and touch it. Then Cynthia would shake the eraser again, and again Cog would reach for it. As we reviewed the videotape, it appeared that Cog and Cynthia were taking turns. But on our development chart we were years away from programming the ability to take turns into Cog. The reality was that Cynthia was doing all the turn-taking, but to an external observer the source of causation was not obvious.

Even though Cynthia was not a naive observer of the robot Cog—she had designed and built large parts of it, after all—she had behaved as though there were more to Cog than there really was. This is not at all unusual, as we will see in chapter 5. Cynthia had engaged the robot at a level that it was not able to operate by itself. She had filled in the behavioral details so that the game of turn-taking with the eraser worked out. But she had done it subconsciously. She had picked up on the dynamics of what Cog could do and embedded them in a more elaborate setting, and Cog had been able to perform at a higher level than its design so far called for.

This got Cynthia thinking about social interaction and how much of what we do with each other can be subconscious, how mothers lead their infants as Cynthia had lead Cog, and how we are often in a dynamic with another person, sometimes leading, sometimes following. Cynthia decided, for her Ph.D., to build a robot that could engage in social interactions.

So through the ministrations of Cynthia Breazeal, Cog begat Kismet.

Many other graduate students helped at many times, but overall Cynthia was the chief architect of the whole process. The final system, pictured in figure 3, had a face that was not quite human. Its eyes were bigger than they should be—we found that people reacted to it as if it were a child, talking in exaggerated tones to it. It had ears that moved, like a dog, but also eyebrows and lips. Kismet was just a head, but it did have a neck so that it could crane its head forward and swing it from side to side. Its eyes could saccade, just like human eyes, and so it was possible for people to understand what it was paying attention to, just as they understand other people, and dogs and cats.

Kismet had a set of fifteen computers controlling it. They formed an equal-opportunity network for operating systems. Some ran a real-time operating system called QNX, some ran Linux, one ran Windows NT, and a few ran an operating system that I had written. The network grew over time as Cynthia was working on Kismet and was very nonuniform in the ways computers talked to each other. The real beauty of this was that there was clearly no one computer in control. Different computers moved different parts of the face and eyes, different computers received visual and audio inputs. There was no place that everything came together and no place from which everything was disseminated for action. Kismet truly was a distributed control system with no central control.

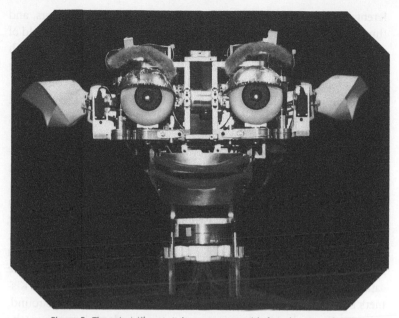

Figure 3. The robot Kismet. It has two eyes with foveal cameras behind the humanlike eyeballs. Two wide-angle, or peripheral, cameras are hidden where its nose would be. Microphones are in its ears. A gyroscope sits in the middle of its head. These are the sensors that Kismet has. Its actuators let it move its neck around three different axes and its eyes to independently scan left to right and together up and down. Kismet also has the ability to make facial expressions—its jaw can open and close, and four separate motors control its lips. Each eyebrow and ear are driven by another two separate motors.

The way computers had been added to Kismet over time enforced that constraint, but it was not one that scared Cynthia. She had previously worked on Attila (see chapter 3) and was used to enticing globally consistent behavior out of completely distributed asynchronous, simple little computational processes.

To make Kismet a sociable robot there had to be many independent systems put into place. A very important part of Kismet is its visual attention system. Cynthia developed this system jointly with Brian Scassellati, another graduate student working on Cog. Kismet saccades toward whatever it is paying attention to. It pays attention to three different sorts of things: moving things, things with saturated colors, and things with skin color. It weighs up these three features across its field of view and then looks in that direction. How much weight it gives to each one depends on other parts of Kismet. If it happens to be lonely, then the visual attention system gives much higher weight to skin-colored regions of an image. If Kismet is bored, it gives higher weight to saturated colors. If Kismet looks at one thing too long, then it gets habituated to it, and something else is more likely to become more salient.

Other behaviors can take over from the pure attention system and direct Kismet's eyes. If Kismet is paying attention to something and it moves, then Kismet's eyes will follow it in smooth pursuit. And Kismet uses its gyroscope to compensate for head motion, as just like a human it has a vestibular-ocular reflex. Kismet finds people's eyes and during turn-taking conversation will make and break eye contact appropriately.

But what does it mean for Kismet to be lonely or bored? Kismet has a set of internal drives that over time get larger and larger unless they are satiated. As these drives get larger, they release certain behaviors. If Kismet's boredom drive is high, it might start deliberately looking around, saccading from place to place in a systematic way. At the same time, the weighting its attention system has on saturated colors will be high. So if, as it searches, Kismet happens to see some bright colors in its periphery, its attention system will direct it to saccade directly toward them. To an external observer it looks like Kismet was looking for a toy and has found one. In a sense, that is what happened, but notice that the search behavior never really finds the toy. It just provides the attention system with widely varied images, and the attention system serendipitously finds the toy. Kismet has a search behavior that never knows when it finds what it is looking for. The overall behavior emerges from the interactions of simpler behaviors, mediated through the world. When Kismet's drives are satiated, by finding a toy for instance, they are reduced markedly, and then once again start to increase over time.

Kismet has an auditory system that not only hears that an utterance is being made but also looks for various prosodic markers in the speech of people talking to it. Universally across just about all human cultures mothers convey four basic signals to their very young children through the pitch patterns in their voice rather than through the actual words they are saying. The pitch variations in someone's voice are known as its *prosody.* Human infants can detect *approval, prohibition, attention-getting,* and *soothing* through prosodic patterns. Kismet, too, can listen to voices and extract these emotional messages from human speech. People in my research group come from a variety of places and we have tested Kismet in English, French, German, Russian, and Indonesian.

These prosodic signals can affect Kismet's mood. Kismet's mood or emotional state is a combination of three variables, its *valence,* its *arousal,* and its *stance.* Kismet's valence is a measure of its happiness, its arousal is a measure of how tired versus how stimulated it is, and its stance is how open it will be to new stimuli. Depending on its current emotional state, prosodic cues, or motion cues, or anything else it can sense, will move it in different directions within its three-dimensional emotional space.

In English we have common names for many places in this emotional space. If Kismet is highly aroused but has neutral stance and valence, then we would say it is surprised. Something coming very close to Kismet's face will arouse it, and if it started out with all three parameters neutral, then it will become surprised. If, however, it was already some-what unhappy, with negative valence, then such arousal might lead to fear or anger, depending on whether its stance was open or closed at the time.

Kismet's internal emotions affect which of its behaviors get activated. But they also get expressed in its face and its voice. It displays its emotional state by the set of its eyebrows, its lips, and its ears. Also, it puts prosody into its own voice, just as it can hear the prosody in the voices of people.

What Kismet cannot do is actually understand what is said to it. Nor can it say anything meaningful. It turns out that neither of these restrictions seems to be much of an impediment to good conversation. Kismet hears only that people are speaking and the prosody in their voices. Kismet utters English phonemes, but it does not understand what it is saying and does not know how to make the phonemes or syllables string together in any meaningful way. Rather, Kismet has the basic mechanisms of turn-taking, with pauses, gaze shifts, and the filling in of awkward silences when its partner fails to speak.

For weeks on end during the spring of 2000, Cynthia Breazeal brought naive subjects into the Kismet lab on the ninth floor of the MIT

AI Lab building. What made them naive was that they knew nothing about robots, and in particular nothing about Kismet. Some of these people were friends of Cynthia's roommate or friends of other graduate students, or teenagers from local schools where the teachers had requested a tour of our laboratory. Cynthia would sit them down in front of the robot Kismet and tell them, "Speak to the robot."

No help beyond that. "Speak to the robot." People were left to their own devices to figure out what that meant and how to speak to the bodiless head that is pictured in figure 3.

Most people were soon able to strike up a conversation with Kismet, even though Kismet only babbles nonsense syllables[5] and does not understand a word anyone says to it.

Cynthia videotaped all the interactions and made many quantitative measurements from them. These measurements show that people are able to direct its visual attention toward the object, that Kismet is able to recognize socially communicated reinforcement from people, that it is able to communicate its internal state to humans, and that it is able to engage in joint regulation of social interactions with humans.

But it is also fascinating just to watch the tapes. Most people know when it is their turn to speak—they really understand that they are participating in a conversation with the robot, albeit a content-free conversation. However, the fact that it is content-free worries some people, and they really do not know what to say to Kismet. It sometimes prompts them repeatedly, and the people do not say much at all. Other people, however, really get into the spirit and chat away to Kismet for minutes on end. They give it emotional cues, and they understand its emotional responses.

From the tapes that I have seen, perhaps the most engaging interchange is one between Ritchie and Kismet. Ritchie was a natural who chatted away to Kismet for twenty-five minutes without getting bored. At one point he says, "I want to show you something. I want to show you this watch my girlfriend gave me." Kismet dutifully looks at the watch on Ritchie's left wrist. To naive observers it seems as though Kismet is understanding the words that Ritchie says. For all we know, Ritchie may

5. More recent work has been concentrating on language acquisition for Kismet, and the robot is now able to say some English words that it has learned.

have thought that it understood his words. But of course Kismet did not. Rather, Kismet picked up on other social cues that Ritchie was giving it subconsciously. He brought his left wrist into the center of Kismet's field of view, a few inches below his face where Kismet was foveated. As he did so, he reached up with his right hand, extended his index finger, and tapped the face of his watch. This visual cue was very strong and got Kismet's visual attention system to saccade to the motion. Kismet looked at the watch as Ritchie spoke. He had its attention, so he stopped moving his finger. Before long, Kismet's attention system decided that Ritchie's face was more interesting than a small patch of motionless skin color. It looked back at his eyes just as Ritchie was done speaking, and took its turn to speak. A completely natural interchange.

There is nothing really in Kismet which is qualitatively different from the mechanisms in Genghis. In the Appendix we go through every detail of Genghis's control program. Kismet's is much too large to fit in this book, but it is the same sort of thing. Little tiny pieces of mechanism that together make a wondrous artificial creature. Genghis could walk around the world like an insect. Kismet can interact with people like a human. Kismet acts like it is alive.

Kismet is not HAL, but HAL was not Kismet either. Kismet gets at the essence of humanity and provides that in a way that ordinary people can interact with it. It has the basis for understandable intelligent behavior, to be built on top of its sociable substrate. HAL was all intellect, but a cold, hard, calculating intellect. It could never be your friend in any meaningful way, and its intellect, as improbable and impossible as it might be in that disembodied form, made it other. HAL could not be understood. Kismet cannot be misunderstood.

Further Reading

Asimov, I. 1950. *I, Robot*. New York: Gnome Press.

Ballard, D., M. Hayhoe, and J. Pelz. 1995. "Memory Representation in Natural Tasks." *Journal of Cognitive Neuroscience* 7(1): 66–80.

Ballard, D., et al. 1997. "Deictic Codes for the Embodiment of Cognition." *Behavioral and Brain Sciences* 20(4): 723–42.

Breazeal, C. L. 2001. *Designing Sociable Robots*. Cambridge, Mass.: MIT Press.

Flesh and Machines

Clark, A. C. 1968. *2001: A Space Odyssey.* New York: New American Library.

Johnson, M. 1987. *The Body in the Mind: The Bodily Basis of Meaning, Imagination, and Reason.* Chicago: University of Chicago Press.

Lakoff, G. 1987. *Women, Fire, and Dangerous Things: What Categories Reveal About the Mind.* Chicago: University of Chicago Press.

Lakoff, G., and M. Johnson. 1999. *Philosophy in the Flesh: The Embodied Mind and Its Challenge to Western Philosophy.* New York: Basic Books.

Menzel, P., and F. D'Aluiso. 2000. *Robo sapiens.* Cambridge, Mass.: MIT Press.

Yarbus, A. 1967. *Eye Movements and Vision.* New York: Plenum Press.

5.   Machines to Live With

In the year 1000 there were a few learned people in Europe, mostly restricted to Ireland, which maintained traditions of learning through the Dark Ages. The monks in the abbeys were engaged in preserving old knowledge, and when possible trying to generate new commentaries and perhaps even a little new idea here or there. They spent their resources on copying older texts. Clearly they were the information technology leaders of the day. Suppose in the year 1000 one of them had been asked to speculate on the future of information technology in the year 2000. Now I admit that this thought game is a little facetious, as probably our learned monk would not have had much in the way of a conception of technology per se, nor any inkling that technologies change over time. But excusing that, what would our monk have been able to say about the technology of his trade 1,000 years hence?

If our monk had been particularly savvy, he might have predicted that by the year 2000 we would have much more finely filed feathers to write with, and a wider variety of colors of ink, and perhaps, just perhaps, faster methods of stretching lambskins to write on. Failing the gift of clairvoyance (and, come to think of it, who better to have that gift?), our hapless monk would have failed to predict the replacement of vellum by paper after a 1,200-year technological journey from China by way of the Muslim world. This technological innovation was of course what ultimately broke the monopoly of the monasteries on manuscripts and written communication. Our monk would have failed to predict the printing press, the personal typewriter, the word-processing computer, the Internet, and the World Wide Web. There's not a chance that he would have had a conceptual inkling of any of these innovations. Of course, if you go back as recently as 1985, hardly anyone in the world save Ted Nelson and his followers would have predicted the existence of the World Wide Web by the year 2000, and not even they were so bold as to think that it would become the major engine of economic growth that it has become.

The point of this anecdote is that speculation on the future is inherently dangerous and doomed to failure. Now this observation does not stop many writers from extrapolating from the kernel of an idea to what might happen during the next century or so. In later chapters I will specifically argue with some of the better-known pundits, such as Hans Moravec and Ray Kurzweil, about our coming robotic/cyber future.

Our monk in Ireland in 1000 would have been able to make a pretty safe prediction that nothing much was going to change technologically by the year 1010, or most likely even by the year 1050. And he would have been right. We live in a time of very fast technological change and development, if for no other reason than that more than half of all the scientists and engineers in the whole history of the world are still actively working now. We get new disruptive technologies coming along with ever increasing speed.

Disruptive technologies are those that fundamentally change some rules of the social games we live with. Napster is an easy example. It is turning the whole model of music distribution on its head. And remember that our music distribution model was a venerable old model that has been with us for more than half a century now. Disruptive technologies have no respect for age-old traditions and practices. And we can expect them to be plopping themselves down in our lives with increasing frequency.

Thus, I think it is foolish, arrogant, and unwise to try to predict the future very far at all. And anyone who tries is doomed to fail. So I will not try to predict the future of our lives with robots beyond just a handful of years. At least not in this chapter. Later, I will speculate on some trends and where there are likely to be new disruptive technologies. But in this chapter I want to take the new technologies that I have already introduced earlier in the book and speculate on how they might well change our lives in the next few years. It turns out that these technologies are already creeping into our daily lives, so we can get some idea of where they are headed, at least in the very near future.

Intelligent Entities

Artificial intelligence technologies have crept into our lives without our realizing it.

Automatic translation from one language to another was an early target of AI. It was a nice domain to show that a computer program could really understand language, because it was a language-based task that did not require the machine to do anything physical or have a large database to access. Both of these were hard to accomplish in the early days of computers. A program that took an input sentence on punched cards in one language and output a sentence on a printer in another language was an attractive domain to work in. Furthermore, it was a very attractive area for government funding, either to translate foreign intelligence information automatically or, as in the case of Canada, to provide an easy way for copies of documents to be produced in multiple official languages. Translating languages turned out to be harder than people originally expected. It was not sufficient to substitute one word such as *main* for "hand" and to follow the rules of grammar in the two languages to fix the word order. "Give me a hand" is idiomatic, and a direct word-for-word French translation seems best restricted to use in a prosthetics factory. Even in less complex language contexts, a detailed understanding of culture and customs can be important for accurate translation.

People have worked on the problem for forty years now, and indeed there has been progress. In my own house we use an Internet site to translate instructions for our housekeeper. We do not speak Portuguese and she does not speak English. We have to keep our sentences reasonably straightforward and declarative, and we translate to Portuguese and back to English again to make sure the correct sense is being kept in Portuguese. It is a workable system, however, and the AI programs that do this use a large amount of knowledge to ensure accuracy over impressively large domains.

The same technologies are creeping into the front ends for Web search engines. In the early days of the Web, say four or five years ago, searches were largely keyword-based, and we tended to get thousands of matches to our keywords, most of which were entirely irrelevant. In the modern world of the Web we now often type our queries as English sentences, and the search engine derives semantic meaning from those words and comes up with a more restricted but usually more helpful set of responses. Often, there are a lot of off-base matches, but in my own experience with these systems, I usually find that at least one of the initial suggested documents gets me the answer that I want.

But language translation and understanding is just one part of artificial intelligence. Those of us who travel by air a lot are finding it less and less pleasant. Airports are becoming more and more crowded, and the roads to and from them are becoming more congested. This trend will only continue. With the increasing number of passengers there are more flights, and so we often get held by air traffic control at our destination even before we have left the ground. Once we get to our final airport, however, it is relatively rare that we have to wait for a gate at which the airplane can park. Most gates are now scheduled by AI systems that are constantly updating the assignments based on where all the other flights are, which passengers have to change to what other flight, what ground equipment is available where, and the status of the new crew for the airplane in question.

When we apply for a mortgage, especially on the Internet or by phone, chances are that we will be evaluated for suitability by an artificial intelligence program, a neural network. The neural network will have been trained on thousands of cases of both good mortgages and bad mortgages. From that experience it has built up a mathematical partition function which, given all the parameters of a new applicant (e.g., salary, time in job, marital status, number of children, etc.), will classify that applicant as good or bad. Although it has not stored all the previous examples explicitly, essentially it is trying to ascertain, from summarized past experience, whether the new applicant is most like a good-risk borrower or a bad-risk borrower.

These examples of artificial intelligence technology are not intelligent entities. They do not have an ongoing existence that is tied into the flow of time. They are not situated. Rather they are procedures that are applied to a current set of data and produce a result, rather like a square-root procedure that is given 9 and returns 3, or is given 700,569 and returns 837.

Recently, however, we have started to see real entities, with longer-term existence—tenuous, but longer-term—show up in our everyday life. These are only very weak entities, but they are the forerunners of those that are soon to be released into the same ecology that we all inhabit.

The paper clip that pops up when we use Microsoft Word is an artificial intelligence application developed by Eric Horovitz and others, based on their Ph.D. work at Stanford. The paper clip often makes some pretty intelligent and surprising guesses about what we are trying to do as our African savanna brain tries to manipulate our fingers into generating beautiful and heartfelt manuscripts.

The video games that we or our children play almost all have an AI engine at the center, turning internal agents into worthy opponents for us in their virtual world. These agents do not need to "see" as we see, for the programmers can make shortcuts so that they have direct access to the state of the whole world that we humans are forced to infer from the visual display we see on the screen. But even grading them down for that cheat, these agents are often much better than we are. And when their skill level is turned all the way up, they can usually beat us. It is not just in chess that the machines are better, but in all the action video games that continually pour from programming sweatshops throughout the world.

When we watch recent movies that have animated crowd scenes, we are often watching intelligent agents, each with graphical avatar, and each interacting with its neighbors. Sometimes the avatars are animals on the African savanna. Sometimes they are ants, hordes of ants. And sometimes they are birds flocking together. The underlying agents each had a very brief life during the animation phase of the movie production, and each was simulated individually, with a simulated clock ticking away as the agents noted the behaviors of their neighbors and adjusted their own behaviors in response.

In the mid-to-late nineties there was a craze, especially among young girls, for small toys with three buttons and an LCD screen not more that 2 centimeters across. The originals were called Tamagotchis, but there were many other versions built in the Far East to capitalize on this craze. A Tamagotchi started out as an infant personality on that tiny screen. It needed care and feeding and nurturing, accomplished by pressing the buttons and selecting appropriate virtual actions from icon-based menus. The Tamagotchi required attention to various needs to keep it in homeostasis. While in homeostasis it grew and developed for weeks and weeks if the owner was sufficiently attentive. But woe to the owner who did not carry their pet with them at all times. If it was not attended to when it demanded attention, things could go badly awry, and the entity within the little oval-shaped pendant could die. In Japan that was it for the toy. Once dead, it was gone and could not be revived. In the United States there was a way to reset the toy and start over again. "You let your pet die? Oh, OK, here's a new one, just like the old one."

In late 1998 there was a hot new toy that captivated first the United States and then the rest of the West and Asia. Known as the Furby, it was developed by Tiger Electronics. After seeing the prototypes at Toy Fair in New York in February 1998, Hasbro, the world's second-largest toy-maker, decided to buy Tiger. It turned out to be a good investment, as Furby was a real hit, one of the biggest new hits in the toy trade in recent times.

Furby was the first embodied entity with an ongoing life of its own to make it into our households. Furby was a robot, though admittedly with just minimal actuation capabilities. Furby was a little troll-like doll, each doll unique, that could open and close its eyes and mouth, and rock up and down. It had a wide range of prerecorded sounds that it played with some randomization, so that it did not ever appear to be in exactly the same state that the owner might have seen it in before. It had a handful of sensors distributed around its body, including a microphone that was sensitive to loud noises. This let the program running inside get some crude idea of how it was being played with. It also had an infrared communication link in its forehead, and when two Furbys were brought face-to-face, they could communicate with each other, engaging in turn-taking conversation and in other synchronized activities.

A Furby was often, but not always, responsive to the way people played with it. It would get into certain modes, or be playing certain "games." It certainly seemed to have a personality, and just as the outward appearance as to exact coloring differed from Furby to Furby, so did the personalities. This was accomplished by setting a few bits randomly inside the toy, but the outward manifestation was different personality types expressed through tendency to sleep, and ease of arousal when someone played with the toy. Furthermore, each Furby changed perceptibly over time, changing both its behavior and the words it tended to speak. Over time, more and more English crept into its vocabulary. The advertising literature for Furby gave the impression that it learned, and that it learned English from listening to its owner speaking to it. This is not what really happened—there was an internal clock measuring how long it had been activated by playing. The longer its play life, the more English words were activated. This behavior and the advertising suggestion was sufficient to convince many owners that it really did learn English from them. It was also sufficient to get it banned from being brought to work at some of the United States government security agencies. There was a worry that Furby would overhear sensitive information and regurgitate it like a parrot at some later time. This fear was totally unfounded.

A few years ago Dr. T. Doi of Sony Corporation set up the D21 laboratory (now the Digital Creature Laboratory) within Sony. The first task he assigned his engineers was to reproduce the capabilities of the robot Genghis, the robot we explored in detail in chapter 3. Dr. Doi hired one of my students, Juan Velásquez, to work on the emotional system for a new doglike robot they were building. This robot was a descendant of Genghis but now had four legs rather than six, and each of those legs was articulated rather than the sticklike legs of Genghis. They were very impressive electromechanical devices, and the Sony engineers were able to make them move in lifelike ways. Juan was charged with giving them an internal life through an emotional model.

In 1999, Sony announced their new robotic dog, AIBO. Through clever marketing and limited supplies they were able to sell out a few thousand of them in twenty minutes in Japan and a little longer in the United States. Very impressive for a $2,500 "toy" that did not do much. AIBO is a four-legged dog with a movable head and a wagging tail. It has a vision system built into it and a powerful processor. The dog has some simple moods and is able to sit, walk, and chase balls. It is like a mechanical pet, though not very affectionate. I think they would have turned out a lot better if Juan had gone back to help them some more.

The Sony AIBOs were largely bought by early adopters—those people who are very motivated to have the newest high-technology gadget before anyone else. Often, it was the parent rather than the child who drove the purchase. There is a way for owners to program their dogs, and the AIBOs have been very popular with some hobbyists. Perhaps the most interesting thing, however, is the way that many of the nonprogrammers view their AIBOs.

Here is an extract from one of the main AIBO owner sites on the Internet, concerning whether AIBO has face-recognition software.

> Sony states that Aibo does not have facial recognition, but that is a feature they are considering for the future. Aibo does seem to recognize people and, particularly, his owners. Most owners will tell you that their dog knows them by sight. It is probably true that Sony did not put specific facial recognition software into Aibo, but nevertheless, his visual system and AI is somehow able to detect and learn new shapes. It seems logical that this would include a shape he sees a lot, like his owner's face.

Of course, AIBO cannot recognize faces and Sony has stated that, but the owners like to think that their AIBO can recognize them. So they have come up with a rationale about how AIBO probably is able to recognize faces after all. Especially their owner's face.

Likewise for speech recognition.

> Sony states that Aibo does not have speech recognition, but like facial recognition, this is a feature that they are considering for the future. Aibo is much like a real dog, in that he cannot understand English, Japanese or other human languages. But like the facial and shape recognition, he is able to connect basic sounds with desired actions if properly trained. Several owners have reported training their dog to respond to certain commands and also to commands from one side or the other (right or left) since he does have stereo hearing.

Again the owners are desperate for their dogs to understand them, and despite Sony's protests to the contrary, they have rationalized that AIBOs can after all understand enough about voices to learn commands. Both face and voice understanding are projections by the owners onto their robotic dogs. This should not surprise us. All of us probably project a little too much intelligence on our flesh-and-blood dogs. It partly explains our level of affection for them. Sometimes we project too much intelligence on our children too. So it is natural that we should do the same for our robot companions, to overanthropomorphize them a little, to make them more like us. At least in our minds, and that after all, is all that counts.

Toys like AIBO are the forerunners of intelligent entities that will come to populate our world more and more. But what are the real issues that drive the design of such toys?

The Life of a Toy

In 2000 the new toys for Christmas were interactive dolls. Some referred to these as "Furbys on steroids," for they took the capabilities of Furbys and pushed them to the next level. I was very much involved in the development of one of these toys, My Real Baby, which was marketed by Hasbro.

Around 1995, Colin Angle, Chuck Rosenberg, and I, at iRobot, had built a robot with a face. We called it IT, perhaps for Interactive Technology. In some ways IT was the precursor of Kismet, although we were much less driven by wanting to build a psychologically accurate creature. IT had eyebrows, movable "eyes" with infrared sensors embedded in them, a set of movable lips, and a neck that could swivel around. I programmed a simple emotional system for the robot, and a few facial expressions that could be blended together to give IT the ability to express its internal state. Chuck cobbled together a few simple behaviors, such as smiling for a flash camera (after the photo was taken) and a withdrawal response if people got too close to it.

This test rig convinced Colin and me that it would be possible to make expressive toys at a practical price. It turned out that we did not yet know enough about toys.

We got diverted for a while by some investors who knew even less about toys than we did, but we eventually settled on building an expressive baby doll. We recruited Chi Won, a master mechanical engineer, from within the company, and in late 1996 and early 1997 the three of us built Bit, or Baby IT.

Bit had a passive body, much like a conventional rag doll. Its head, however, was something else. Chi developed a remarkable set of plastic cams and levers, driven by five cheap motors. Attached to this box of mechanism was a rubber face. Chi spent weeks getting the right mechanical properties for this artificial skin, so that as the levers moved, the face deformed in a lifelike way. Bit could smile, frown, screw up its face as if about to cry, look surprised with raised eyebrows, look scared, and appear angry. All with only five cheap motors. We were very proud of ourselves. Chi got some of the plastic parts out of other toys, not because he could not design and build them himself, but because we wanted to be sure our toy worked with low-quality toy parts. We were even prouder of ourselves.

Colin and I worked on the electronics and software. We used a microprocessor that we knew cost less than $10 if purchased in quantity, and we programmed it in a language I had developed for behavior-based robots back in the Genghis days. We programmed in an emotional model, so that Bit could have the many different emotions that its face was capable of expressing. One of iRobot's engineers recorded sounds made by his baby boy; and we burned that into ROM (read-only memory) on the processor, so that we could play back snippets at any time. We chose some very cheap sensors, which we embedded in Bit's body and under its clothes. Then Colin and I wrote carefully crafted code to interpret what the changing readings from these sensors could tell us about how Bit was being played with. A small cage containing a ball bearing gave the software clues about whether Bit was being gently rocked, or bounced up and down, or even if it was being held upside down. Magnetic sensors in Bit's mouth told us whether its magnetically tipped bottle was in its mouth. Light sensors under its outer layer of garments let us determine whether it was being hugged or tickled, and a microphone let us distinguish between the high-pitched voices of young children and those of adults.

We ended up with a doll that was very lifelike, much more than any toy on the market at that time. We had tried to model the behavior of a real baby as much as possible. If the baby was upset, it would stay upset until someone soothed it or it finally fell asleep after minutes and minutes of heartrending crying and fussing. If Bit was abused in any way— for instance, by being swung upside down—it got very upset. If it was upset and someone bounced it on their knee, it got more upset, but if the same thing happened when it was happy, it got more and more excited, giggling and laughing, until eventually it got overtired and started to get upset. If it was hungry, it would stay hungry until it was fed. It acted a lot like a real baby.

Since this was before the arrival of Furbys, we really thought we had something new, and in fact Bit was much more advanced than a Furby.

Colin and I did know enough about toys to realize that we would not be able to manufacture, market, and sell such a toy ourselves. In previous years I had made the rounds of Japanese toy companies with a friend from Canada, Takashi Gomi. We had been trying to sell them on the idea of behavior-based robotic toys, but without a prototype.

Now Colin and I had a prototype toy to show. In fact, we had two— the second was a talking, self-steering ball, using much of same technology that we had developed for the doll. So we went on the road to try to sell our toys to U.S. toy companies. Now we started to get a real education.

The first thing we discovered was that there is an expected price point for a doll, which happened to be much less than just the cost of the parts in our doll, let alone the manufacturing, packaging, shipping, distribution, marketing, supply-chain profit, and anything that might come to us. The percentages that went to all these things astounded us, and when we realized that the cost of the parts should be only about 6 percent of the sale price, we knew we were in trouble. For all our pride in the low-cost mechanisms, sensors, and processing that we had developed, we were still a factor of twenty away from where we needed to be. How were we ever going to get past that hurdle? A factor-of-twenty price reduction is not something that comes easily.

The second thing we discovered was that having a lot of new features was not necessarily a good thing. When we showed our prototype to toy companies, we often got enthusiastic initial responses. Then people would start counting up the new features. Five, six, ten, depending on what they saw as being interesting. And then the message that astounded us. "We can't possibly go with more than one of these new features. That's all you need to sell a new doll. And besides, you only have thirty seconds in a TV commercial. You can't get across more than one new idea in thirty seconds, so anything else is wasted!"

I made a trip to Taipei in the middle of 1997 to hang out with some people from very small toy companies so that I could learn some of the ropes in building toys cheaply. One of the people I visited was an Englishman in his early sixties who had been building simple toys, some with speech chips embedded in them, for many years. I followed him around for a couple of days while he did deals. I never saw him sit at a desk—he would hold court at a conference room table surrounded by toys that he was currently selling all over the world. No paper, no calculator, no spreadsheets. Just him and his brain. People would come to the office, chat with him a few minutes, and then be off. Some were bringing prototype parts for his own version of a Tamagotchi. He had hired someone in Hong Kong to write the software for him for a few thousand dollars. He chatted with me off and on about how big Tamagotchis really were going to be. I did not know. But I discussed the topic with him anyway. On the third day I was with him, we chatted some more, and then he told me he had made his decision. He would build 8 million of them. Eight million!

As I sat astounded by this, wondering how much money my enthusiasm was going to cost him, he started calling suppliers. First, he called the supplier for the custom LCD screens. These were the most difficult items to get, and so he needed to know the supply possibilities there first. I only heard one side of the conversation, but as I understood it, he was being told how many he could get on what dates. No notes, no records; he just chatted for a while. The next person he called was going to make the plastic cases. "What do you mean, four cents each? I can't afford more than three and a half cents a piece!" A little yelling and shouting and soon he had a deal, and a delivery schedule for 8 million plastic cases at 3 1/2 cents each. And so on down the list (all in his head) of components he needed. Then he started calling the people he supplied around the world. France: "I'll be delivering two hundred thousand units to you on October eighth, then another two hundred thousand a month later." Not a spreadsheet in sight. I could only imagine the conniptions a Harvard Business School M.B.A. would be having at this point. It was certainly a different world from the one I was used to.

Between the things we were learning at toy companies in the United States and my foray into Far East manufacturing, we were getting a little disheartened. On the other hand, we had made some good contacts in Taiwan and learned who the players were in building cheap chips. We knew what sort of scruffy little microprocessors we should be aiming at. We had learned how to get unpackaged silicon flakes with 20¢ processors on them manufactured for us in Hsin-Chu Industrial Park a couple hours south of Taipei. We had learned how these chips needed to be hand-carried, never shipped, through Hong Kong and into southern China where they would get bonded directly onto a printed circuit board—no little chip packages making them look like multilegged insects. That could add pennies to the cost of manufacturing the product. And, we had learned how to get the "plush," the soft part of the toy, manufactured elsewhere and integrated with the electronics in the relatively high-tech parts of China.

This is how electronics-based toys are manufactured in today's world. Trying any other scheme would lead to a noncompetitive product. We were going to have to live with the logistics of Far East manufacture or give up our dreams of building behavior-based robot toys for the masses.

By late 1998 we had managed to come to an agreement with Hasbro. Our company, iRobot, would suggest ideas for toys and develop prototypes, and they would handle the manufacturing and distribution. We started out suggesting many tens of new concepts, but before long we settled down on building a version of Bit that would come in at the right price. Hitting the price point we did required a lot of adjustment on our part, and a lot of new technical innovations. Hasbro had to adjust their thinking quite a bit too, as even with radical cost-cutting measures My Real Baby ended up being more expensive than conventional wisdom said the market could bear. Fortunately the one thing that Colin and I did not learn on our toy odyessy was conventional wisdom.

We started work on My Real Baby in early 1999. We had to get our software system running on tiny processors to be made in Hsin-Chu Park. We had to get a team of software developers willing to use that software system, targeted to cheap chips with only a few hundred bytes of RAM when they were used to writing large programs on machines with tens of megabytes of RAM. We had to find a way to preserve the functionality of Bit's face while reducing from five motors to one. We had to develop new sensors that cost less than a penny each. And we had to turn all this into a product the public would understand and appreciate. For the first time, we were going to have a mass-market robot, not one that lived in a research lab. For the first time, our robots were going to have to interact with countless thousands of real people in ordinary homes, not graduate students interested in esoteric aspects of human psychology.

My Real Baby turned out to be quite different from Bit in the way it acted. MRB does not try to act like an actual real baby in the way that Bit did. We tried to make MRB a rich play experience for children rather than a chore for them to tend to. MRB has an internal emotional model, and it sometimes gets upset, happy, hungry, or even virtually damp. But unlike Bit, MRB does not insist that, for instance, you feed it when it is hungry. If you do, it soon gets satisfied, perhaps asking for "more" if you do not feed it enough. And then it might burp if you put it on your shoulder and pat it on the back. But if you decide not to feed it, MRB eventually gets over its hunger and is willing to play whatever game the child wants to play.

So while My Real Baby is an entity in the world with wants and desires and an ongoing existence, it is not quite the same sort of entity as a real animal. That was a conscious decision made on the part of the designers, wanting to let children have the initiative in play patterns so that they could exercise their imaginations.

By the way, the good thing about a virtually damp diaper is that you only have to virtually change it, not really change it. MRB is quite satisfied if you open up the diaper and then put the exact same one back on.

When My Real Baby hit the stores just before Christmas 2000, we found out that there was perhaps some value in conventional wisdom after all. The television ads could not get across the excitement of the doll that everyone who bought one reported. There were too many new features, and the advertisements were too busy with details, getting beat out by simpler dolls with only one trick up their sleeves. We sold more robot dolls than any other robot ever in existence, but it was not quite the megahit we had been hoping for. However, by early 2001 all the major toy companies had decided that robot toys were what was going to be the big seller for the next few years. We had a new walking dinosaur, a robot with attitude, ready to hit the market by the middle of 2002, with more and more robots on the way.

Computers in Our Lives

Robots are following the same path that computers took but are lagging, generally speaking, by twenty or twenty-five years. Just as we saw the first few computers in research laboratories, then their acceptance into industry, behind closed doors and away from the lives of ordinary people, so we have seen the same steps with robots. Next we saw computers' entry into our homes as toys, first as simple video games like Pong, and then the first few telecommuters started to use them for real work. Finally the killer applications of e-mail, instant messaging, and the World Wide Web made computers ubiquitous in our everyday home lives.

If the early parallels between computers and robots hold up, we can expect to see such killer applications for robots in the next ten to fifteen years, and by the year 2020 robots will be pervasive in our lives.

Robots in Our Lives

In 1958, Joe Engelberger developed the Unimate, a hydraulically powered robot that was a big, dumb arm. It could be set up to go through a series of motions, again and again, moving from place to place. It might open a gripper to pick something up at one place and again to put it down at another place. Or it might activate a spot welding gun attached to its "hand" at various stops. It was big, strong, and incredibly stiff, so that it carried out its actions with great repeatability—every time it went through its motions, it arrived at exactly the same spot. Engelberger was a fantastic salesman and managed to convince Detroit car manufacturers that using these robots on their production lines would be cost-effective. He was right, and soon his and competing robots flourished in the United States, Japan, and Europe. Lots of robot manufacturers sprang up, all building hydraulically powered arms.

A few research laboratories started to build electrically powered arms during the sixties. At the Stanford Artificial Intelligence Laboratory a pair of so-called Stanford arms were built. They were good for research but too idiosyncratic for use in the real world. In 1971, Victor Scheinman, a mechanical engineering graduate student at SAIL came to spend a year at the MIT Artificial Intelligence Laboratory. He developed a small, human-sized electrically powered robot arm. It appeared much more humanlike than previous arms with an identifiable shoulder, elbow, and wrist.

By the time I landed at SAIL in 1977, Victor had created a company to market the Vicarm, but it soon got bought out by Engelberger's company, which had become a division of Westinghouse. There an arm twice the size of the original but with the same design as the MIT version was developed. Eventually that business got bought by Kawasaki Heavy Industries, and now these PUMA robots, as they are known, are the most common electric arms in the world. Many other companies have also gotten into the business, both Japanese and European.

These large hydraulic and electric arms are now indispensable to modern automobile manufacture. They are dangerous machines. They hardly sense their world at all, and they move about with tremendous velocity and force. People are not allowed near industrial robots for their own safety. Smaller electric robots are used in chip manufacture, and there people are banned because we are too dirty and will contaminate the silicon wafers unless we are dressed head to toe in special clean garments that hold in our dandruff and other bodily detritus. The result was the same. Robots were closeted away from ordinary people, and only the priesthood could get close to them to interact with them. This is precisely the situation we had with computers even into the eighties. They were hidden away from ordinary people, and only somewhat nerdy specialists got to interact with them in person.

Unlike our early computers, however, robots did have another side. Hollywood had turned them into machines that did interact with real people. Sometimes it was to subjugate the masses, but other times they were just more sentient beings in the diorama of life, such as in the *Star Wars* trilogy. The general public knew about robots from these depictions but thought of them as something far in the future. And the general public knew about them as mindless machines that sometimes threatened jobs—although one would be hard put to find anyone in the world who could say that they were thrown out of work and replaced by a robot.

Colin Angle, Helen Greiner, and I knew we could not build robots that lived up to either the positive or negative Hollywood expectations. But we did realize that it was important to build robotic-based toys and get them into the mass market. They would sneak into people's homes, and they would lead to much lower-cost robots that could be used for all sorts of applications. Guessing those applications, finding the "killer app," is of course the generic problem of the new technological world we inhabit.

Apart from a toy, why would you want a robot in your home? They are probably not going to be much good for storing recipes or doing your taxes—we can agree that our computers are pretty good at doing that for us. Robots in our homes will be useful if they can do some physical work for us, or if their mobility somehow helps us in some way.

Perhaps the most common request for a home robot is one that cleans the floors. The Swedish company Electrolux (and its United States subsidiary Eureka), the German company Karcher, the British company

Dyson, the Japanese company Minolta, and my own company, iRobot, all have prototype home-cleaning robots that they intend to bring to market in the early part of this decade. The big question is, how well will they clean, and at what cost? For actual consumers, of course, an additional question is, how easy are they to use?

Building a cleaning robot for the home has long been a fascination of robotics researchers. Most of them have not concentrated on the cleaning mechanism so much; rather, they have concentrated on coverage. How to get the robot to reliably find its way around the house and cover all of the floor. The general thought had been that this would require the robot to have a map of the house and to know where it was at all times, so that it would know where it had been and where to go next. In recent years the companies developing home-cleaning robots have largely given up on this requirement. It is too hard to do in any cost-effective way. Ideas for this ranged from having beacons set up around the house so that the robot could triangulate where it was at all times, to having an onboard three-dimensional vision system that built a complete and precise geometric model of the world, to having the robot keep track of precisely how far it had moved and in what direction, so that it always knew exactly where it was.

None of these three ideas was practical in the eighties and is still not today. Requiring a consumer to install beacons around the house is certainly not a very practical idea. The beacons could be passive objects that are detected by the robot, like bar codes posted on the walls or small circuit elements that do not require a battery but respond to radio waves sent by the cleaning robot as it shuffles around the house. Or, the beacons could be active elements plugged into wall outlets or with batteries that need to be changed every so often. Unless all our houses get built with some sort of beacon system installed, as we expect electricity and water to be installed at the time of construction, we are not likely to see home robots relying on beacon systems. And even if new houses get built with the beacons installed, that still leaves the vast majority of our housing stock without them. When electricity and electric lights were introduced late in the nineteenth century, most houses were retrofitted over a twenty-year period to have electricity in every room. Electricity was very compelling, and it remains to be seen just how compelling robots will become.

A three-dimensional vision system, despite thirty-five years of work, is still only a partial success. There are some prototype systems that produce a three-dimensional model of where stuff is, and where stuff is not, but the results are similar to building an adobe model of a house interior with fist-sized-or-bigger lumps of clay. The modeled floor is rough, and the shapes of the furniture and doorways make it difficult for a human to interpret what they are, let alone a computer vision system. Even with the three-dimensional model it is difficult to know how to make the robot clean the floor and to know where it has cleaned.

The last idea, keeping track of how far the robot has moved, is called odometry, just as the mileage gauge in a car is called an odometer. There are two problems with this for a small indoor robot. The first is that it is impossible to know the exact orientation of the robot, and even a small error, say 1 degree, turns into a 5-centimeter error once the robot has traveled 3 meters. So moving across a small room five or six times will get the robot confused about where it is by as much as a full body size—this makes it impossible to track which area is clean and which is not. The second problem is that the robot will not be able to track how far it has moved very well, as its wheels will slip variable amounts depending on the exact properties of the floor surface. Any robot will move differently on a tiled floor, a wooden floor, or a carpeted floor. For carpet the direction of motion is also critical—moving with the nap or against the nap, the direction of the fibers, results in about a 10 percent difference in distance traveled for the same number of wheel turns. That means our cleaning robot will be uncertain about a body length in travel as it crosses a small room.

In order to be reasonably priced and reasonably simple to install, we need to give up on the idea of our robots knowing exactly where they are at all times. At least we must give up on the idea that they know where they are precisely enough to know where they have cleaned and what still needs to be cleaned.

In the late eighties a few of us started to think about a new strategy. Let the robot bumble around your house, or a room, and let it clean as it goes. It may end up cleaning the same spot many times, but eventually it will randomly cover all the spots and then the target area will be clean.

My group at the MIT AI Lab built a prototype vacuum-cleaning robot to do just this. Sozzie had two extra features.

First, in the tube that all the dirt was sucked through, we mounted a laser at right angles to the direction of flow, and measured how much light was making it across the pipe. More light meant that the vacuum mechanism was finding less dirt to suck up, while less light meant that the vacuum was successfully extracting lots of dirt from the floor surface. This information was used to modulate how much the robot would head off in a straight line versus how much it would muddle around in the current general area. In that way Sozzie tended to spend more time in the dirty areas of the house and scoot off if things were fairly clean already.

Second, Sozzie knew how to recharge itself. Its charging station had an infrared beacon that Sozzie could detect. Sozzie monitored its battery level and if it was low and it happened to pass by and notice the beacon, it would give up cleaning and head straight for the recharging station. This is a little like Walter's tortoises that we described in chapter 2. Sozzie's opportunistic recharging strategy did not completely guarantee that it would never get stranded somewhere in the house with run-down batteries. The threshold as to when it would choose to recharge, should it see the beacon, could be adjusted and would trade off the likelihood of getting stranded against how much cleaning work it would do between recharges.

Based on observations about how well Sozzie worked, I at one time extended the idea of a cleaning robot to an ecology of cleaning robots all living and working in a house together. Sozzie sucked up dirt only within the perimeter of its body. A person cleaning a house with a vacuum cleaner has a nozzle at the end of a long pipe, and they push that nozzle under the couch and force it into the corners of the rooms. A Sozzie-like robot might never get to all those nooks and crannies. It seemed that a smaller robot could come in handy for those sorts of places, but it would have even less chance of finding its way back to a charging station where the owner could occasionally empty its vacuum bag.

The straightforward engineering approach to this problem is to increase the expense of the small cleaning robot and force all the complicated mechanism into the smaller space. Another approach is to change the way one thinks about what a robot should do, and see where that leads. This is most likely the source of the true innovations that will come and will change the fundamental way we think of robots. This is much like what happened when people stopped thinking about computers as machines to do mathematical computations and started thinking about them as data-storage and communication machines.

In the case of home-cleaning robots this led me to think about having one or more very small robots that are able to clean the hard-to-reach parts of all our homes. I envision these as small hockey-puck-sized robots with small legs that they use to slowly, slowly drag themselves around. Let's call them *pucksters* for now. Like the robot Genghis, pucksters would rely on their physical interaction with the world to navigate. They would keep moving until they hit a wall, then drag themselves along the wall-floor interface, until they found the corner of a room. As they went, they would pick up little pieces of dirt, perhaps electrostatically, and store them in their belly. At corners of rooms, which they would sense by feeling vertical surfaces both in front of them and to one side, they would stay for a while, picking up lots of dirt as that is where it would tend to accumulate. If they ran into a chair leg and tried to follow that as a wall, they would soon realize that it was not a wall and head off again looking for another candidate. Likewise, when a puckster was following along a wall, every so often it would get wanderlust and choose another direction.

Eventually one or more of these pucksters would get to all the corners of a house. If made small, they would be very robust and could survive a fall down the stairs. Perhaps they could be programmed to notice such a fall and just sit there, as though stunned, right at the bottom of the stairs. When the homeowner saw it the next day, perhaps, they would carry it upstairs again and toss it down on the floor anywhere they chose. By making the pucksters slow we can reduce the amount of energy they need to a level that they could harvest themselves through solar cells, mounted on both the top and the bottom of their bodies. Top and bottom, so that if they tumble over they can continue to work—why should our little pucksters need to worry about which way is up on their body. These little robots would be very slow indeed, perhaps moving only a few meters per day, carefully picking away at the dirt that they imagine surrounding them. Whether there is dirt at any particular location or not is irrelevant. They act the same in any situation, patiently dabbing their small legs about, fishing for little dirt particles. Ultimately, however, they would get the job done. If they were made cheap, consumers would get the choice of how clean the corners of their home were by occasionally buying another puckster or two until things seemed sufficiently clean.

But what are these pucksters to do with their bellyfuls of dirt? They could rely on the ecology within the house. They could listen for when the big vacuum cleaner was active, then abandon whatever they were doing and rush to an open bright area of the room, attracted by the more intense light. As in a suicide run, they could use up all their energy resources, then spill their guts, pouring out all their collected dirt into a pile on the floor, and lastly drag themselves away from their dump site to sit and recharge. The big vacuum cleaner would treat them just like any other obstacle and scoot around them, but ultimately it would randomly stumble across the pile of dirt the puckster had left and suck it up to take it back to its recharging base.

An ecology of robots could thus cooperate to clean your house. But notice that they do not have to explicitly cooperate. Rather, they cue their behavior off each other without ever sending any messages to each other. The puckster hears the big robot and does its dash in response. The big robot treats the puckster as an obstacle and independently happens across the dirt pile it has created.

This is a very organic sort of solution to the housecleaning problem. It is not a top-down engineered solution where all contingencies are accounted for and planned around. Rather, the house gets cleaned by an emergent set of interacting behaviors, driven by robots that have no explicit understanding of what is going on, nor of how to accommodate gross breakdowns in the expected ecology. It is a balance, robust over a wide range of conditions, which the designers have conspired to make work. That conspiracy allows very simple robots, and therefore very cheap robots, to work together to get a complex task done.

The housecleaning robots that are on the market or about to be on the market are not as radical as the ones I have outlined above. They are, however, at odds with the earlier thoughts on how a robot *had* to work in order to be effective. All the companies have converged on an understanding that a random cleaning pattern will suffice to clean the bulk of our modern houses.

Each of the robots has a fairly simple interface. The robot is placed on the floor and switched on. It goes about cleaning the floor until picked up or until some time period is up. The user does not need to give it a map of the house, and the robot does not try to make one. It bumbles around, perhaps with some special behaviors that make it follow walls when it bumps into one, to try to pick up all the dirt along the edge of a room. There is no certainty that any particular spot on the floor will be covered, although the probability that it will be gets closer and closer to certainty the longer the cleaning robot operates. All of the robots have some sort of collision-avoidance system built in, but none of them currently can return by themselves to a recharging station. When their time is up, they simply switch themselves off wherever they are.

So this then is the immediate future of life in our homes. Small robots that we grab from a recharging station, twist a knob, plop down on the floor while we walk away to another part of the house. Perhaps we close the kitchen doors after us to confine the robot so it will stay there cleaning away. Dumb, simple robots. Robots that move about in our house with our initiation but without our intervention. They will be new *almost life-forms* that coinhabit our houses. The models coming out in the next year or two range from far too expensive for the mass market to cheap enough to be bought by many people as a whim purchase. There may be more than one model that people will want simultaneously in their houses. Besides the floor model, there will probably be small robots to clean the kitchen countertops, and another one specifically for dining room tables. All of them will be robots that their users set to work and then forget. All that will be necessary is an impulse to clean the floor, or the countertop, or the tabletop. There will not be any need to worry about it for longer than a couple of seconds, and the task will be done.

This is rather like the home-heating situation for those of us who live in the U.S. Northeast. Sometime in the fall we set our thermostats, perhaps with different temperatures for different times of the day, switch on our oil heater, and forget about operating it until the we switch it off in the spring—except when we pay the monthly oil bill. The oil company delivers oil to our house autonomously, and since they monitor our average usage and adjust for the recent weather they make sure the tank does not run dry. We do not need to gather fuel, light a fire, clean the hearth, nor worry about smoke in our houses. It all happens automatically, buried in a boiler room somewhere down in our basement.

The new cleaning robots will ultimately make housecleaning as automatic. However, it will probably not be quite as out of our consciousness because our hordes of cleaning robots will be moving about in our living space. Perhaps this will be comforting, but perhaps it will turn out to be distracting. Before very long, before the middle of the first decade of our new millennium, perhaps these cleaning robots will be forced by the marketplace to become rather more discreet in how they share our houses with us. The technology already exists for them to find their ways back to their charging stations. Before too long the main floor-cleaning robot might hide in the closet and venture out only in the middle of the day when it senses that no one is around the house. It will clean the floors and scuttle back to its hiding spot while we are gone. Unless we come home early and catch it out and about. It will notice someone is around and abort its mission for the day. If there are people around every day, it will get more and more frustrated and eventually venture out in desperation even when people are about.

Over time these cleaning robots will get cheaper and cheaper, just as the computers that power our wristwatches have gotten far cheaper than anyone dared imagine only twenty years ago. Just ten years from now it may turn out that little shelf-dusting robots will be cheap enough to be thought of as disposable. You will buy a ten-pack or two of them at the store and put a little thimble-sized robot on every windowsill and mantelpiece. They will consume light energy and slowly and continuously dust their assigned space, trapped on one side by the window or the wall and on the other by a ledge that they sense and avoid. Our houses will become menageries of little cleaning robots. A symbiosis will develop between people and the artificial creatures whose only role in life is to keep the house clean.

Common, Common, Common

When I was a teenager, a laser was an exotic device even harder to find than a computer. I spent tens and maybe hundreds of hours trying to build a laser (unsuccessfully) before I ever saw one. That only happened in a physics lab when I got to college. Now I am not at all sure how many of them there are in my house. There are certainly more than ten compact disc players in my house. Some are in laptop computers, others in desktop computers (six people means more than six computers these days), and a stereo player or two on each of the three floors of the house. All of them have a laser in them. Then there are the laser pointers that I stuff into my pocket whenever I am going off to give a talk somewhere. And perhaps there is a laser or two in each of the VCRs (again at least one on each floor of the house) as part of the motor servo mechanism. Perhaps not. The point is that I do not know. And I do not care. What was an exotic device of great mystery and excitement just thirty years ago is now a common household object too numerous to bother cataloging.

Cleaning robots will become common in our houses, but so will other sorts of robots. What might they be used for?

Some of our domestic chores became quite automated during the twentieth century. Washing and drying clothes has been reduced to collecting the laundry and putting it in the appropriate machine. That collection is still a chore. But on the output side there is still ironing, hanging, and folding the dry clothes and putting them away in their appropriate closets. The ironing component has been partly automated in two ways. The first is through the adoption of new materials for clothes that dry in a wrinkle-free way. The second is an increase in sloth, wherein we are more willing to wear unpressed clothes, perhaps fooled into a false sense of security by the wrinkle-free claims on the labels.

Washing dishes has become a lot easier with the advent of the dishwasher, but it still has some drawbacks. We still need to pick up all the plates from the table, scrape them off and put them in the dishwasher. And there are lots of large pots and pans that need to be scrubbed by hand, along with the wooden-handled knives, and other things that are not "dishwasher safe."

Before you decide that automating these remaining tasks would be frivolous, remember to think through whether having a washing machine or a dishwasher is frivolous. It certainly is in many ways, but it is a convenience we have all come to depend on and all think that it leads us to a more fulfilling life in the amount of time it frees up for us. Automating the rest of the job will have a similar sort of effect. Once it is cheap to do, we will all wonder how we could have gotten by without it.

Now we get to the question of how these tasks should be automated.

The first thing to think about is that perhaps a robot is not the ideal solution. After all, if instead of building a dishwashing machine, we had waited until we could build a robot that could wash dishes at the kitchen sink, we would still be washing all our dishes by hand.

Handling clothes to put them in the washing machine or dryer seems perhaps the easiest of the tasks we have been discussing. There is no need for accurate handling and no need for precise control over the folds in the clothes. Grabbing the clothes in pretty crude ways will probably work. The trick is to get them into the right machine, with the wide possible variety of door arrangements, then getting the right amount of washing powder or liquid into the washing machine, and getting the clothes out of the machine when they are all plastered against the sides of the revolving drum after spin-drying. All of these tasks can probably be done by robots in the near term, and the only question is whether a robot that can do them can be made cheaply enough for people to feel it is worthwhile.

The ironing is a little trickier. The best approach to this will probably turn out to be something more like a dishwasher, i.e., a special-purpose ironing machine rather than a robot wielding an iron. Ironing involves delicate handling of floppy and large pieces of material. The sensing and skill necessary to do this is considerable, which partially explains why so few of us are good at it. Just about anyone can pick up the laundry and throw it into a washing machine, but ironing requires real skill. Just the visual sensing problem alone is far beyond the capabilities that our computer vision systems have. For these reasons I do not think we will see a robotic solution to ironing in the near future.

What about cleaning up the dinner table? Before another decade is out it is probably going to be possible to have a robot that picks up the dishes from the table and takes them to the kitchen. But it is rather unlikely that it will be able to place each dish in the dishwasher, as that is a rather dexterous operation. Dishwashers have been designed to be filled by people, and the designers optimized how dishes fit into them, knowing that dexterous people would be packing them.

Perhaps the whole table-setting and -clearing problem will require a completely new way of thinking about the problem. Perhaps in the future our dining room tables will also be the place that we store all our dishes when we are not using them, in a large container under the tabletop surface. When we want to set the table, small robotic arms, not un-like the ones in a jukebox, will bring the required dishes and cutlery out onto the place settings. As each course is finished, the table and its little robot arms could grab the plates and devour them into the large internal volume underneath. With direct water hookups into our dining rooms, much as we have them now in our kitchens, the dining room table could also be the dishwasher. It would wash the dishes after the meal and leave them down in its recesses, waiting for the next meal at which they are needed.

So not all automation in the home is going to fall to what we might traditionally think of as robots. But there will be a lot of places where we will see them.

Let's continue our fantasies on where else we might have robots in our homes in the not-too-distant future. There are a whole collection of ideas that rely only on current technologies. Of course, the big question is whether someone will find a way to embody them in a cheap enough, and reliable enough, robot, so that people will want to buy it.

We might soon be confronted every morning by a robotic magnifying vanity mirror. The versions right now sit at the end of an orienting arm, and we can change the mirror's orientation by hand. Before too long it might be augmented to follow us around the bathroom, so that when-ever we are ready to use it, it will already be positioned at the perfect angle.

We might soon see robots that each live on one window in our house and periodically clean it. We might soon see cleaning robots for the bathtub and for the shower recess. We might soon see a robot that can clean toilets. Less likely in the short term, but believable in the longer term is a robot that can make our bed, pick up our dirty clothes from the floor, and place them in the laundry basket. Rather more difficult to fathom right now is the robot that can unload our groceries from the car, including putting them away in the refrigerator. This sort of robot might need to wait for a fundamentally new kitchen organization that facili-tates the functionality in some new way.

Now we can move on to even more frivolous sorts of robots. The one desired by many a sports fan is the one that responds to the verbal com-mand "Bring me a beer." Surprisingly this is not too far away. The tech-nology is virtually at hand to achieve this. The robot needs to be able to navigate around the house but not in a precise way, as with the early ideas for a cleaning robot. It does not need to line up with centimeter accuracy to make sure it precisely covers all of the floor area. Rather, it can move from room to room, avoiding collisions with the furniture using sonar-based navigation. As long as it knows the topological layout of the house—and there is technology around that lets today's research robots very nearly build a good enough map of a house just from wandering around—then it can find its way from the living room to the kitchen and back. For retrieving the beer from the fridge one can imagine two assists to the robot. A small grip place could be added to the bottom of the fridge door so that the robot could open it without having to accommodate the variations from fridge to fridge. Then inside the fridge a special beer can dispenser could be placed on the shelves, stocked with beer cans, probably lying on their sides.

Although the grip place sounds like a reasonable addition, somehow the special beer can dispenser seems a little disappointing. First, it means that the cans can only be retrieved in the order they were placed in there. So much for being able to say, "Hey, robot, get me a beer, a Diet Coke, and two ginger ales." But worse, it is a special-purpose solution, so a robot that can get a beer is unlikely to be able to get your slippers, or running shoes, or a clean pair of underwear. I think the truth is that we may have to wait a few decades for that level of convenience.

There are going to be more and more robots in our homes. Pretty soon we will stop bothering to count them. They will be a new class of entity, moving about under their own free will, doing their tasks as they decide they need to be done. The ecology of our homes will be visibly more complex than it is today. Just as our houses with their refrigerators, washing machines, dishwashers, stereos, televisions, and computers would look to someone from a century ago sort of like a house but with a whole lot of weird stuff littered about, so too will the houses a century from now look a little strange to us.

6.   Where Am I?

E arly in the year 2000, I brought home a robot lawn-mowing machine. There are a handful of companies that sell them now, and
they range in price from a few hundred dollars to two or three
thousand.

My machine was at the cheaper end, but as far as I can tell, not a whole
lot separates the capabilities of any of the robot mowers. Some of the
manufacturers celebrate the fact that their lawn mowers are robots, while
others go out of their way to avoid using the term "robot" at all. I bought
mine from a company that was up front about it. The mower itself was
the size of a small wheelbarrow and looked vaguely beetle-shaped. It did

not cut the grass and store it somewhere; rather, it mulched it into very tiny pieces and left it on the ground as it mowed.

The method all these robot mowers use to achieve coverage of the lawn is a variation on the theme that is being used by the home floor-cleaning robots. They randomly traverse the required area; statistically they eventually cover all the spots if they are left to operate long enough.

In order for the robots to know what area to cover, they need to be told somehow. All the manufacturers have chosen a variation on a simple idea. A wire is pegged down around the perimeter of the area to be mowed and left permanently installed. Whenever it is time to mow the lawn, the user needs to switch on a unit that sends a signal down this wire, emitting low-power radio waves broadcasting a unique pattern. The robot mower can use this wire in two ways. First, it can follow it around the perimeter to get a cleanly done edge of the grass. Then when it is randomly covering the interior region it measures the radio waves and notices when it is at the boundary. It backs away and changes direction slightly, randomly covering the desired area.

I have a fairly large lawn, so I pegged off about a third of it and then pressed the big GO button on the robot. I was impressed with how well it worked. It went around the edge easily, then started cutting away at the interior area.

My beloved spouse, Janet, was not as impressed. Instead of the straight parallel markings left by the manual mower, there were lines at all angles across the lawn. Worse, after two hours, when the batteries had run down to the point that the robot sat beeping for assistance, there were still lots of random tufts of grass, some bigger than the size of a dinner plate, that had not been mowed. I manually drove the robot over to a power outlet and plugged it in to recharge for the rest of the day. The next morning I set it going again. By the time it finished that day, I was long gone to work. When I came home and plugged it in overnight, there were just half a dozen or so fist-sized tufts of grass that had not been mowed. It seemed that manually driving the robot over these was more prudent than rolling the dice for a third time, since the cost to me of trying again was actually fairly high. I had to set the robot going in the morning. That was pretty easy. But then I had to worry about retrieving the robot in the evening, perhaps in the dark if I got home late, and manually drive it to be recharged.

The lines on the lawn disappeared after a couple of days and it looked as nice as if had it been mowed manually. But the amount of work the robot was supposed to save was worrisome. Mowing that section of lawn manually took only about fifteen minutes. The total amount of time it took me to tend to the robot was perhaps a little more, and was spread out over multiple days. This was not a fire-and-forget robot. Admittedly, I did not get as dirty as I do mowing the lawn manually, but the experience was not quite satisfactory.

What were the root causes of the failure of this technology to deliver a worthwhile experience? There were two fundamental problems. Batteries and navigation.

If the robot had been able to carry enough batteries to operate for six hours, say, then it would have been able to do a pretty good job and I would only have had to interact with it twice—once to set it going, and once to drive it back to the garden shed when done. The problem with batteries is that they have not followed Moore's law, and there is no reason to expect that their performance will radically improve anytime soon. Batteries are better than they were twenty or forty years ago, but not fantastically better. The bigger a battery's capacity, the heavier it is, and so the bigger the capacity battery one needs to power the mechanism that drives it around. It is almost, but not quite, a losing battle in trying to increase the battery life of a mobile robot (or an electric car for that matter). The longer the running time, the heavier and more cumbersome the robot must get. Six hours—three times more than were provided—would have been enough to mow my patch of lawn. But that section was only a third of my lawn. Eighteen hours is what I would have needed. Instead of being barely liftable out of the trunk of my car, I would have had a robot that required a crane or a forklift to unload once I got it home from the store. Clearly batteries by themselves are not the right solution.

The other shortcoming of my lawn-mowing robot was that it did not know how to navigate by itself. When its batteries were running down, it simply stopped in its tracks and beeped plaintively, waiting for someone to come rescue it and drive it to a recharging station.

Navigation, especially outdoors, is a hard problem, and one that does not have any good low-cost general-purpose solutions. Indoors the walls and corridors and furniture provide a lot of constraint. Sonar works well, unlike outdoors where its reflectivity from bushes and shrubs is very variable. We have robots that can navigate around a house well enough to get to a recharging station, though not well enough to ensure complete cleaning coverage.

The global position satellites (GPS) can be used to get an approximate position outdoors, and this is what is used in car navigation systems. GPS suffers from having many dead spots, especially when near buildings, so the car navigation systems interpolate between good readings. They use the knowledge they have of the road structure, along with readings from accelerometers, to infer where the car could possibly be as it moves about. This gives the driver a continuous update of estimated position that is usually accurate to within a few meters. That is more than enough accuracy for driving about, especially when there is a human in the loop looking at a map and relating it to all the visual cues of roadways that they can see simply by looking out of the window. All the automatic driving systems that researchers have built likewise need to take visual cues from the world to update their GPS and accelerometer position estimates in order to be able to drive without running off the road. So our lawn-mowing robot would probably need vision to augment a GPS-based navigation system. Of course just the cost of the GPS system alone would have added perhaps 50 percent to the cost of my robot. Adding vision would double or triple that again. But the real problem is that we do not yet have vision algorithms that would work reliably enough over the wide variation of foliage and terrain that robot lawn mowers would encounter in suburbia.

So it seems that for at least the next ten years or so our low-cost outdoor robots will not have good navigation abilities. It may be twenty or thirty years before this problem is cracked. It looks like we are still going to need a human to drive the robot back to the recharging station. The good news is that the human does not have to be there to do it.

Figure 4. The remote presence robot, iRobot-LE. Six wheels drive the robot around on flat surfaces and two more swing down from the front to help it up and down stairs. They can also push downward so that the robot can stand up in "prairie dog" mode. The stalk in the middle of the robot can lift up to see things well above tabletop level. In the head is a pan-tilt camera. On top of the head is a sonar scanner that gives the robot a continuous update on the obstacles that are around it.

Remote Presence

In August 2000, I made a quick trip to Tokyo and Taipei to demonstrate a new commercial robot, the iRobot-LE, made by my company. It was about to be announced in *Wired* magazine and I wanted to give some Far Eastern companies a heads-up on what was about to happen.

I arrived at a nondescript office building not far from Tokyo Station, walked down a corridor, and entered a scene from *Total Recall*. I was in a stainless steel corridor, stainless steel walls and ceilings, without a mark anywhere. I got to the end of the corridor, another blank stainless steel wall, and it opened in front of me. I entered the most un-Japanese-seeming office complex I have ever seen in Japan—light and airy, with glass and steel everywhere.

About twenty executives were coming to my presentation. I had just a few minutes to plug in my laptop and get my Internet configuration set up to get through all the company's firewalls. There was no time to test anything beyond making sure that I could bring up an iRobot Web page in my browser. The executives trooped in and I gave my slide presentation. Then I clicked on the browser, bore down to a demonstration page that required a password, and suddenly I was transported back to Massachusetts.

On the projected screen for all to see was real-time compressed video coming from a robot in an apartment in Boston. I clicked on a little control widget and panned the camera around. I clicked on a place in the image, and the camera rolled right over to that place. I clicked on a place above a couch and the camera rolled toward that place but then stopped, indicating that it could not get there from here. The camera was on the head of a stalk on a robot like that pictured in figure 4. In Tokyo we could hear the whir of the robot's gears as it moved. Since it was midafternoon in Tokyo, it was two in the morning back in Boston, and in any case the apartment was just for demonstrations like this, so there was no one around. If there had been, I could have chatted with them, and looked them in the eye. I was present remotely. I was "inside" a remote-presence robot halfway around the globe.

Back in October 1998, in Tsukuba Science City in Japan at a robotics conference, I had seen a videotape of a remote-presence experiment. About ten engineers in Japan were gathered around a control station. They were on the telephone to a group of approximately ten other engineers in Pisa, Italy, who were gathered around a mobile robot. With everyone nursing along complex pieces of software, the Japanese researchers were able to give supervisory commands to the Italian robot and have it carry out simple movements. There was much cheering and shouting when everything worked.

Now, less than two years later, remote presence required no engineers, no cheering, and no shouting. From anywhere on the planet I was able to project myself into a commercially available robot and have it do my will. Best of all, I needed no special software in my location, just a standard Web browser.

I was not driving the robot as I would with a joystick. Rather, I was giving it high-level *supervisory* commands, telling it what I wanted it to do, and it took care of the details. When I told it to go to a certain place, it would take care of locally avoiding any obstacles, using a sonar scanner to get an up-to-date measure of where stuff was that might be in the way. When I told it to look up at tabletop level, it extended its neck so that it could see things that high. When I told it to go up a flight of stairs, it took care of all the details, detecting the side walls with its sonar and worrying about yaw and slip as it slithered upward.

Controlling a robot like this with a joystick would not work very well, as the time lags induced by the Internet would soon confuse anyone trying to do so. Tom Sheridan had shown that years ago at MIT when he was experimenting with human subjects controlling robots to handle nuclear materials. If the lag between command and action and feedback to the person gets longer than about half a second, then a person is unable to keep track of what actions they have commanded. Soon they steer right, say, get no response, so steer right harder. As soon as they see the robot steering hard right, they start steering left, and not seeing an immediate response, steer even harder left. Before they know it the robot is oscillating wildly back and forth, and they are struggling to keep control. If they simply eased off, everything would work out fine. Despite knowing this cognitively, even highly trained operators simply cannot control a system when there is a high *latency*—a long lag time round-trip between command of action and seeing the action take place.

The Internet has latency in it, especially when you are halfway around the world. Commands cannot travel faster than the speed of light, and they may be bounced way up into the sky and back from a geosynchronous satellite. That travel time soon accumulates to a few tenths of a second. Furthermore, the commands and return video images need to go through many switching computers along the way. Every way point adds a little more time, and soon the latency limit is passed and it becomes impossible to joystick-control a robot in any stable sort of way.

That is why remote-presence robots can only be controlled in a supervisory manner. But it turns out to be a feature rather than a bug. By letting the robot handle all the local decisions, there is much less cognitive load on the remote human operator. It is much easier than driving a car, and it is alright to be doing something else at the same time. The robot takes care of its safety and the safety of those around it by itself.

If the lawn-mowing robot I had bought had had remote presence it would have been a much more effective robot. I could have set it going in the morning before heading in to work. A couple of hours later I could have "robotted in" to it and projected myself into its body. Of course, I would have been at work, but via my Web browser I would have felt that I was in the robot, in the same way we feel like we are driving that Formula One car at the video game arcade. Then I could drive it over to the recharging station from my office at work and forget about it until after lunch. Then I could robot in again, and set it back out on the lawn to do the cleanup passes. Perhaps I could drive it back to the garden shed late in the afternoon before it got too dark to see where I was going.

With this new improved lawn-mowing robot the effort that I had to put in at home would go down to zero. I could mow my lawn while on vacation in Australia, for a total of about eight minutes connected to the Internet.

How much bandwidth is needed *out* of the home for this to work? Up until now everyone has been worried about how much bandwidth can go into the home so that the occupants have good Internet and video access. With remote presence, robotting in back home, the equation gets reversed. The robots become Web servers (and indeed the iRobot-LE has onboard a Linux box running a secure Apache Web server).[1] At iRobot we have limited all our designs to require no more bandwidth out of the house than can be commonly achieved with DSL, which is 300,000 bits per second. This is much slower than can be achieved with cable modems, but faster than a telephone modem, or ISDN. One can be sure that the Web economy will push people to higher and higher bandwidths into and out of their homes, but what most people in the United

1. The Apache Web server was first developed at the MIT Artificial Intelligence Laboratory by Robert Tau. He was distressed that the home-brew stock-market tracking system that Mark Torrance, one of our graduate students, had put up was swamping perform-

States can already purchase for a low monthly fee is enough for remote-presence robots. No doubt as more bandwidth becomes available, people will come up with interesting new ideas for remote presence that will make use of that bandwidth.

The Future of Work

Here now is the killer app for robots in the short term. Physical work can be done from any place in the world. The implications of this will be profound on the world's economy.

The first place that such remote work may become very popular is in Japan, and indeed the Japanese government is financing advanced research in this area. Japan has a unique combination of factors that interlock in a way that makes this possibly their only way out of some problems looming over the next few decades. The factors that combine in Japan are resistance to immigration, a very low birth rate, and, thanks to the high standard of health care, a booming population of older people.

The population of Japan is much more ethnically pure than that of any other advanced nation. The population is over 99 percent pure Japanese, with less than 1 percent foreigners. The latter tiny fraction includes the native aboriginal population of the northern islands called the Ainu, all westerners, and five generations of Koreans who have lived, worked, married, and been born in Japan. It is very difficult for a foreigner to become a Japanese citizen, and so the only source of population growth and new workers is from within, rather than through immigration as in North America, Europe, and Australia.

The birth rate in Japan appears to be the lowest in the world (fol-

---

ance of the "legitimate" Web serving we were doing. So Robert started to patch, and patch, the Web server we had, eventually replacing all of it in a cascade of patches. He thought it was "a patchy" Web server, and thence came the name *Apache*. Meanwhile, Mark's stock site got so popular that we had to come up with a plan for him to move it to a private site, where he started StockMaster.com, which later was sold to Red Herring.

lowed by Italy) with a current average of just 1.3 births per woman over her lifetime. This is clearly not enough to sustain the population, and it has been declining since 1950 when the rate was 3.5. Most projections agree that the rate will not get back to the self-sustaining number of 2.0 in Japan any time before the year 2050.

Neither of these two facts would be a problem for Japan if they had low life expectancy, but luckily for individual Japanese, because of the nation's wealth and excellent medical system, they have almost the highest life expectancy in the world. It is over eighty-one for women and above seventy-five for men. This combined with the low birth rate means that Japan has a higher proportion of older people than any other nation. In the year 2001, 24 percent of Japanese were sixty or older while only 20 percent were under twenty. This compares to 11 percent and 29 percent, respectively, for South Korea, another industrialized nation in the area. But Japan's numbers are even more askew compared to other large Asian nations, such as China with 12 percent of their population sixty or older and 31 percent under twenty. Indonesia has 7 percent and 41 percent, respectively.

This high proportion of an older population gets to the crux of Japan's problem. According to Hamid Faruqee and Martin Muhleisen of the International Monetary Fund (IMF), by the year 2025, Japan will have only two working-age people (aged twenty to sixty-four) for every retirement-age person (those sixty-five or older). This means that younger workers will need to support more people than they do currently, and the situation will get worse in Japan for many decades beyond 2025.

To be sure, there are similar trends among the populations in other wealthy nations, but they are not as severe. The IMF predicts 3 workers for every retiree in the United States, Canada, Germany, and the United Kingdom in the same time frame, and roughly 2.5 such workers in France and Italy.

The population of Japan is aging, and there are not enough young people to fill the low-grade jobs. The first place this was noticed was in agriculture, especially because Japanese farms tend to be based on small plots of land dug into hillsides—a land form that does not readily accommodate mechanization. Now the shortfall is becoming noticeable in construction, and more importantly in nursing. As the population gets older, there will be more and more requirements for personal services to help the aged and infirm. Japanese are living longer, but they will have no one to look after them in their old age.

Other wealthy countries are facing similar problems, but they have a different solution. They import cheap labor. In the United States, agriculture is very dependent on immigrant Mexican workers. Even the high-technology industries are dependent on immigrant labor, as was seen in the strong lobbying from Silicon Valley to Congress in 2000 and 2001 to increase the number of H-1 visas. These visas allow highly educated workers to come from countries like India and China, and to work in the (still) booming U.S. high-technology world. In Europe the solution to the labor problem is workers from Turkey, North Africa, and the former Yugoslavia, along with high-tech workers from India and China.

Foreign workers in the Japanese workforce make up only 0.1 percent of the total. In the United States the proportion is 10 percent, and in the European Union it is about 5 percent.

Japan must look elsewhere to solve their labor problem. The government has been looking toward robots. The Ministry of International Trade and Industry (MITI) has been encouraging the development of *friendly robots,* robots that can interact with people in close quarters. The work is proceeding well, but slowly. There is a real concern that autonomous robots will not get good enough quickly enough to meet the needs of Japanese society. To augment the robots' intelligence many people have proposed having a person in the loop, but that does not really solve the labor problem—unless the person in the loop in a nursing home in Japan is not themselves in Japan.

Remote presence provides a way out of the Japanese conundrum. Foreign work, yes, foreign workers, no. And the United States and Western Europe will not be far behind. They too may well turn to the importing of physical work by renting the minds of workers in foreign countries, minds that become the supervisory controllers of robots across the globe.

How We Get There

The future is one thing, but what can be done with remote-presence robots today? The answer is rich and surprising, and it is quite clear how some of the first few steps toward the new world economy will play out. The details beyond that are unpredictable, as this is a disruptive technology that will quickly play havoc with any models that people have. Innovation will come quickly from all sources—from nineteen-year-old college dropouts, from old-economy companies, and from the strangest recesses of humanity, all in totally unexpected and unpredictable ways.

The first question is, what are remote-presence robots good for today, right now? If they have no utility, then people will not buy them, and the economies of scale necessary for changing the way the world works will never get a foothold.

The early adopters are already lining up to buy remote-presence robots for their homes. There are lots of simple uses for them with their current capabilities. It is unlikely that any single one of those uses justifies the financial outlay—it is the sum of the uses that makes them worthwhile.

You are driving to work and just cannot remember whether you turned off the stove as you rushed out with your coffee in hand this morning. It gnaws at you and gnaws at you. Should you turn around and go back to check? Today, if you are close to work, you can continue and as soon as you get there fire up a Web browser (this is probably the first or second thing you do every morning anyway), and robot in to your home remote-presence robot. A sonar-based map that the robot has built of your house comes up in one corner of the screen, while the center is filled with real-time video from the robot's camera. You click on the kitchen in the map and the robot heads there. As it comes into the kitchen, you can see the stove and click on that in the image. The robot heads over toward it, and soon the camera image fills with the door to the oven. You click on an icon of the robot to extend the neck to the highest position and the stovetop comes into view. You click on a pan-tilt icon to swivel the camera to take a close look, and you heave a sigh of relief. The oven is indeed switched off. You click on the GO HOME button and disconnect. Your remote-presence robot takes care of getting itself back to the charging station.

This is today's scenario. You can do this now. In just a couple of years things will get better. In Europe, as 3G telephones come into existence, you will be able to do all this from your cell phone, including seeing the images and clicking on them. When you first get worried about whether you turned the stove off, you'll be able to pull over to the side of the road if you are driving, or just directly control your home robot while continuing your commute if you are a car or train passenger. Unfortunately, 3G phones will take a little longer to get to the United States.

Meanwhile, there will be new generations of home remote-presence robots, and retrofits that have been designed for the current generation of robots. These will put simple manipulators on the robots, and before too long you will be able to switch off the stove when you have accidentally left it on.

Of course, there is a competing technology for this particular task. Soon companies will be offering Internet-ready stoves. With one of those you will be able to directly check and control your stove from your Web browser without a robot intermediary. But, to apply this strategy you will need to replace your stove, which is a once-in-twenty years activity for most people. Likewise, you will need to replace all the other appliances in your house, along with light switches, etc., to get full Internet control. Your pets offer a particular challenge in getting them connected to the Internet, although that will eventually come, as we will see in chapter 10. In the meantime, with a remote-presence robot you will be able to check up on your pets from work and, with the right sort of feed dispenser, even care for them while on vacation. A robot offers a single point of investment that automates your whole house. And you get to take it with you when you move into another unautomated house, whether you are buying or renting your residence.

If you have a second home, then a remote-presence robot will be a way to monitor it from afar. When you hear that a storm has passed through the area, you will be able to fire up your robot and check the house, without having to drive for three hours the next weekend to make sure that all is well for the winter. You will be able to drive around upstairs and downstairs, checking all the windows visually to make sure that things are secure.

Security is also of interest in your primary home. Many people have alarm services. Sometimes at work you get a call from them that the alarm has been triggered at your home. The security company then offers you a choice. They will do nothing, or they will call the police. If you have had the police out to your house often for false alarms, they will be reluctant to go again, or they will start charging you a stiff service fee. With a remote-presence robot in your home you will be able to quickly look around in all the rooms, not just the single room in which you might happen to have a Web camera installed. You will then be able to make a decision on what to do with some confidence, and more likely than not you will not have to race home in the middle of the day to check on the state of your house.

Another form of security is checking in on baby-sitters. With a remote-presence robot in your house you will be able to wander in, robotically, at any time while you are away for the evening, and cheerily greet the baby-sitter and your child. If you are away on business overnight, you will be able to have a conversation with your children as they are getting tucked into bed—not exactly the same as being there in person, but a lot better than a telephone call, especially for very young children. After school your remote-presence robot will provide a little more parental contact for your latchkey children who would otherwise spend two or three hours in the house without any adult supervision.

More nefariously people will use the robots that they can buy today to spy on their neighbors. Many large apartment buildings in New York and Chicago have a TV channel on their cable systems that is a camera in the lobby. This is so that when someone up on the twenty-fifth floor gets buzzed, they can quickly get an image of who it is that is requesting that they be let into the building. It turns out that this channel is the single most watched channel in many of these buildings. People leave it on in the background to watch who is coming and going from the building, who is coming and going with whom, and who is visiting whom. A remote-presence robot lets suburban residents engage in similar amateur anthropological studies.

Researchers, such as graduate student Chris Stauffer at the MIT Artificial Intelligence Laboratory, have developed computer vision programs that can watch a scene and over a period of a few hours learn what is usual and what is unusual. The system observes motion in the scene. It quickly adapts to the waving of trees in the breeze and ignores that motion as not being salient. When watching a busy street, it soon adapts to the cars going by. But if eighteen-wheel trucks are unusual, then it flags one of those whenever it goes by as being a much larger than usual moving object. Or, if watching a house, it will notice when someone comes in or out of the house.

One can guess that before very long there will be a third-party software vendor offering an "I Spy" package for remote-presence robots. The purchaser will leave their house as usual in the morning. But as soon as they get to work, they will drive their home remote-presence robot over to one of the windows, extend the neck, and focus in on Mrs. Smith's house across the road. Later in the morning they are working away when they get an instant message from their home robot: "Something is going on." They switch their browser Web page to their home spy camera, and yes, they were right! Mrs. Smith is getting a new couch!

Now think for a moment about your elderly parents or grandparents. A remote-presence robot installed in their home could be used in a number of ways to extend the time for which it is safe for them to live at home without managed care. The most straightforward is that you or a service would check in with them at an appointed hour every day. Your relatives would give the remote person a go-ahead in response to a phone call to robot in. If your relatives did not answer the call, the remote person would be authorized to robot in in any case—perhaps they will check the house and find that your relatives have gone out for a stroll, in which case they will call in again later. But perhaps there is some serious problem and the person who robots in will find your relatives indisposed and call an emergency service. For a family taking this on themselves there could be a different designated robot driver each day. If everything is all right, then whoever robots in can just have a nice visit with the elderly people. Chat with them, and see how they are looking. If one of them is bedridden, then the roboteer can drive into their bedroom to visit them there, then go downstairs to take a look at some newly blooming flowers that the spouse wants to show off. A variation on the "I Spy" package could alert you at work when things did not seem right at your relatives' house, again through instant messaging, paging, or e-mail. After an unanswered phone call you would robot in to check out the situation with your own eyes and ears.

Looking at the flowers that your grandfather is growing brings up another potential use of remote presence. One of the uses of the Internet—obvious in retrospect, but surprising initially—has been the formation of special-interest groups. These groups discuss everything from Beanie Babies to model trains to porcelain collections. There is a natural desire for people to group together and have discussions of common interest. The recent availability of digital cameras with ports to download directly to home computers has made these discussion groups much richer. The members can exchange photographs of their items of interest. With Web cameras they can have some real-time interaction. With remote-presence robots the arena opens even wider.

In traditional garden clubs members take turns hosting the other members at their house to show off their gardens. All the members tend to be from the same local area, for otherwise travel is a serious problem. With remote-presence robots a garden club can be distributed across the planet. As long as the host of a given day's meeting has a remote-presence robot, all the members can take turns robotting in, and browse and wander through the garden looking at the plants and flowers from the viewpoint they choose, while engaging in discussion with the host and all the other members. The other members will all hear and see exactly what the robot is seeing and hearing, and be able to add their voices to what it says.

Garden clubs are just one possibility. As the Internet has shown, there is a vast imagination and yearning, and people everywhere will quickly come up with novel and unexpected ways to increase social interactions over a wider area than previously possible.

Besides these home uses there are also business uses possible. We already use our robots within my company to replace travel and teleconferencing.

Since our company is spread out over three states, there are lots of engineering meetings that involve teleconferences or videoconferences. Those who have participated in such things know that there are certain limitations, and that the people not in the same room as you are somehow different. They are less connected with what is going on with you and your physically present colleagues. Preverbal communication happens between those in the same room—raised eyebrows or index fingers, nods, and glances. All convey extra levels of interaction that are just not available to those at the other end of the line, be it a telephone line or a video link.

Early in our development of the iRobot-LE a teleconference was about to begin involving engineers in Massachusetts and one in New Hampshire. In the Massachusetts meeting room before the conference call could be placed, in rolled a very early prototype remote-presence robot. "It" was the engineer from New Hampshire. At least, he had robotted in to that robot and was controlling it from his browser in the Granite State. As the meeting proceeded, Todd was able to chat with the other engineers, turn to face them, and draw their attention to him. He was there much more so than when he used the videoconferencing system. Todd could move about, and follow the group when they went to look at a prototype in the next room. Nothing beats being there, unless you can be there without having to travel there. Remote presence gives you the best of both worlds.

The Next Few Years

It will require only a small modification to the larger remote-presence robots that are currently on sale and a minor new attachment to your existing refrigerator door to enable the robots to open that refrigerator at your command. Now when it is time to come home and swing by the supermarket, you can easily check on the contents of your refrigerator to remind yourself of exactly what you need for the evening meal.

Refrigerator doors are easy to open, because there is no latching mechanism—they are magnetically sealed and one just needs to apply a force in the right direction and the door swings open. Internal house doors, even without locks, are a little more difficult to open. Robots in labs have been able to reliably open doors, so it is technically plausible to add that capability to remote-presence robots as a supervisory level command option. Once this is done, it will make it more convenient to drive a remote-presence robot around all areas of your home—you will not have to rely on the internal doors being left open.

If we then add one extra new capability to our remote-presence robots, we literally open up a complete new range of applications for them. We need to let them unlock doors from inside the house. And they need to be able to lock them again. If they can open internal doors, they could open many front doors of houses from the inside, as the unlocking mechanism is tied into the turning of the door handle. Locking those doors is a little harder and requires more dexterity. Likewise, unlocking and locking a dead bolt requires a fair amount of dexterity. I do not think it likely that our remote-presence robots will have that amount of dexterity within the next handful of years, but perhaps they will in a decade.

Therefore, in the immediate future it may be necessary to have a new sort of locking mechanism that can be electrically as well as manually operated. This mechanism would fit exactly in the place of a current generation locking core, so that a homeowner could replace their lock as easily as they can now with a screwdriver. There would be no external power connected to the new electric lock, and no battery. This would ensure simplicity of installation and reliability. Usually the homeowner would operate it manually just as they do now. However, there would be some insulated metal pads, so that a remote-presence robot could drive up, touch the pads, and power an electrical unlock and locking mechanism within the lock. Now the owner of the robot will have the ability to click on a browser window button remotely and their front door will be unlocked.

Why this capability? We have all had the experience of ordering a new piece of furniture, or calling a repairman, where we have been told that the item will be delivered, or repaired, Thursday of next week. The only catch is that the dispatcher does not know at all what time of day the person will be coming, so it is our responsibility to be at home all day next Thursday to let them in when they show up, whether it is at 8 A.M. or at 4 P.M. None of us find this a very satisfying arrangement. With a remote-presence robot that can unlock the front door, all this can change.

When the delivery or repair person gets to your home, you soon learn about it. This happens either through the person calling you or through the remote-presence robot being alerted by the doorbell (or perhaps with a direct radio link from the door button to the robot) and alerting you on your computer at work. Now you connect to the robot, and visually verify that they are who they should be and chat with them as you let them in. If they are delivering something to you, show them exactly where to put it. If they are there to repair or install something you escort them to the place of work. In either case, you are able to follow them around as they work in your house and lock up after they leave.

Now it is true that once you have let in such a person, your robot itself is not going to provide much direct security against them doing something wrong, like rifling through the jewelry drawer and taking everything of value. But then probably your physical presence would not provide much direct security against the same sort of thing. Your physical presence provides a social prohibition of such behavior—you will have direct visual evidence of their misdeeds and that will be sufficient evidence in a report to their employer and to the police. Your presence via a robot will have given you the same sort of direct physical evidence, and by speaking to the person through the robot you will ensure that they will operate under the normal social contracts.

Your remote-presence robot thus frees you from having to wait at home as often as you do now. It also enables the last ten feet for e-commerce, the distance from the delivery truck to inside the front door. So many things that we all buy on e-commerce Web sites get delivered and are either left on our front doorsteps, where they can get stolen, or, instead of being delivered, just a note is left, and we must go and collect them from the delivery company. With a remote-presence robot you can always be virtually at home and ensure that your packages get left inside behind a relocked door.

So now we have the following scenario for someone living in the Northeast of the United States, say. It is late fall and it is time to have your oil-fired furnace serviced for the coming winter season. This involves some relatively simple maintenance: taking out some of the burner components and cleaning them, and potentially bleeding a few pipes. This is not something that any robot will be able to do by itself in the foreseeable future. So it is necessary to get a serviceman to come to your home to do it. Today you are forced to wait at home for the repair person to come to your house. In just a very few short years you will be able to go off to wherever you may wish and then robot back via your remote-presence robot to let the repair person in.

But is that even going to be necessary five to ten years from now? As newer and newer models of remote-presence robots come out, they will have more dexterity added to them. There will be market pressure for this. For instance, suppose you check your stove from your vacation and find that you left it on in the mad panic to get some coffee made and down your gullet in the early morning family rush to the airport. If your remote-presence robot has enough dexterity, controlled by you from your tropical beach house, then it can be used to turn the heating element off; you will be much happier than if the robot is still just a clumsy oaf that can only look around.

Soon you will be wanting it to be able to open and close windows, get more dog or cat food and pour it into the bowl, and take things out of the refrigerator so that you can get a better view of what is really in there. All these dexterous operations will be controlled in a supervisory manner by you via your browser wherever you happen to be running it. The remote-presence robot will not have to be particularly intelligent to do any of these things. That is why I am so confident in my technological predictions here—there is nothing that is beyond the current laboratory state of the art.

So now we get back to the oil furnace repairman you have asked to come to your home to do the five-to-ten-minute servicing job that is required every year. If you already have a dexterous remote-presence robot in your house, why not let him robot into that and do the job from his home office? He does not have to get into his van and drive ten or fifteen minutes to your house. Or, even better, he does not have to leave his hometown in South Asia and travel to your home in the United States or Europe. Certain classes of physical work can now be done from anywhere.

And this brings us back to the general-purpose housekeeping robots that we discussed in the previous chapter. We will soon have low-cost specialized robots that do more tasks for us in our houses. It seems unlikely that we will have a general-purpose robot any time in the next couple of decades that can iron clothes, load and unload the dishwasher, unload our groceries and pack them away, or really clean the bathroom. All these tasks are things that many people hire domestic help to do for them, especially if they are a two-career family with demanding jobs.

In the foreseeable future, such families will be able to hire people who do not live in their own country to do such housework as supervisory controllers of a remote-presence robot in the family home. The brains of people in poorer countries will be hired to control the physical-labor robots, the remote-presence robots, in richer countries. The good thing about this is that the persons in that poorer country will not be doing dirty, tiring work themselves. It will be relatively high-paying and desirable work for many places where the economy is poor. Furthermore, it will provide work in those places with poor economies where no other work might be available. One can imagine work centers being set up in countries where there is a surplus of labor but not yet much Internet infrastructure extending to people's homes. High-bandwidth Internet would be available in the work center, and "menial laborers" would come to work at computer workstations in a relatively pleasant work environment. Not only could household work be done this way; the same people could provide the extra control necessary for the lawn-mowing robot, and other remote-presence robots that we have not yet even started to think about.

While it is easy to generate pages of ideas about how remote presence is going to flourish, the details will no doubt be surprising as they unfold, just as the details of the Internet have turned out differently from just about everyone's speculation from ten or twenty years ago. One element of the equation of which we can be sure, however, is sex.

What is one of the biggest uses of Internet instant messaging? Flirting and online sex. What is the biggest bandwidth consumer of streamed video on the Internet? Web cameras providing live pornography, often with instant messaging or telephone backchatting. What will one of the early big drivers be for remote presence? Guess. . . . Actually, if you guess well and with the right innovation, you may well make a lot of money. Someone is going to.

Further Reading

Goldberg, K., and R. Siegwart. 2001. *Beyond Webcams: An Introduction to Online Robots*. Cambridge, Mass.: MIT Press.

We might notice our dog and make eye contact with it, cocking our head in imitation of the way it cocks its head as it looks at us. We might engage in play with our dog, waiting for it to pounce and then making some reciprocating motion. We pay attention to our dog when it greets us at the door when we get home. We would be devastated, tearful, and emotionally wrought if we saw it seriously harmed. After it has died, we might for years afterward occasionally reflect with an inner smile on some of our interactions with it.

We do not generally engage inanimate objects in the same way. But naive people we bring off the street to interact with Kismet do engage in some forms of these activities with Kismet. Kismet and Cog have an "aliveness" about them. A few years ago, in the early days of the Cog project, Professor Sherry Turkle visited our lab. Sherry has always been something of a critic of the claims of artificial intelligence research. In recent years she has been studying how children have sharpened their distinctions between what is alive and what is not alive as they have been confronted with toys that speak, respond to stimuli, and seem to have an ongoing internal life. After visiting our lab she wrote in her book *Life on the Screen:*

> Cog "noticed" me soon after I entered its room. Its head turned to follow me and I was embarrassed to note that this made me happy. I found myself competing with another visitor for its attention. At one point, I felt sure that Cog's eyes had "caught" my own. My visit left me shaken—not by anything that Cog was able to accomplish but by my own reaction to "him." For years whenever I had heard Rodney Brooks speak about his robotic "creatures," I had always been careful to mentally put quotation marks around the word. But now, with Cog, I had found the quotation marks had disappeared. Despite myself and despite my continuing skepticism about this research project, I had behaved as though in the presence of another being.

Despite not wanting to be affected in this way, Turkle found herself responding involuntarily. The students who work with Kismet report the same sort of effect. They know what Kismet is capable of, and they know exactly how Kismet does what it does, since they are the ones who have programmed it. Often they are working in the same room as Kismet, and it is looking around the room or interacting with someone else. Kismet sinks into their unconsciousness as they are buried in the details of their work. Occasionally another student working on Kismet remotely might reinitialize Kismet's visual system. When this happens, Kismet goes through a calibration procedure so that its sensors can tell it which direction its eyes are pointing. To do this it scans both eyes over as far right as they can go, then as far left, then up, then down. During this time Kismet is not expressing any emotions, nor is it engaging in any of its humanlike behaviors. The students in the lab usually find this unsettling. This lifelike object that their consciousness is somehow aware of but is suppressing in detail, suddenly becomes nonlifelike.

We would most likely be unsettled if our dog suddenly started acting in the way that Kismet does during its recalibration. That sort of action is just not something we think of coming from an animate being. Clowns and mimes make use of this effect by eliminating the natural *joie de vivre* of a human and act as a machine. Should we accord Kismet the status of a being? Or is Kismet simply a machine that occasionally happens to be switched on. We can treat Kismet as something of a thought piece, because if today's Kismet itself does not qualify as worthy of having the status of a being, one can ask three further questions. First, is it possible even in principle for a machine to have the status of a being? Second, if so, then what would it have to have beyond what Kismet currently has in order to qualify as a being? Thirdly, even if we granted a machine status as a being, whatever that might mean, what sort of status should we, could we, or would we grant it?

Levels of Beingness

Before trying to answer these questions directly for Kismet or one of its descendants, let us return to our dog and then some other animals.

We are certainly all quite comfortable granting dogs *beingness*. They are living creatures, as are mice, elephants, and cats. We might debate, however, whether they are conscious. They certainly all seem to have feelings. Our dogs display fear, excitement, and contentment. It is more debatable whether they show gratitude. Most people would be willing to extend these attributes to all mammals, although mice and rats certainly seem less emotionally complex than dogs or horses. Nevertheless, we can all see a trapped house mouse exhibiting a fearful response, quivering, and breathing, and looking around for an escape route. Their fear seems visceral. We can relate to it—it is the same sort of fear we know that we feel in desperate situations. But visceral fear is not the same as reason. It is an open and often debated question whether even chimpanzees can reason. The answer to the question of whether dogs, rats, or mice can reason tends to fall along the lines of pet owners saying yes and scientists saying no.

These animals all have some aspects of beingness, though certainly not all that humans have, even forgetting for the moment their lack of syntax and technology. With this beingness we humans grant these animals certain respect and rights. While there is debate on the role of animals, and mammals in particular, for scientific experiments, almost no one argues that cruelty to animals for cruelty's sake is not immoral. In fact, there seems to be a correlation between those who treat humans with psychopathic cruelty and childhood histories of extreme cruelty to animals.

Now consider reptiles. When you quickly approach a lizard, it acts startled and runs away. It acts as if it is afraid, but are we willing to say that the lizard is really afraid? It does not have quite the same sort of reactions that a mammal has when it is afraid. A lizard has steerable eyes, and we can get some cues from it about its direction of gaze, so we can tell when it is looking in the direction of a source of danger. We may be able to see its rib cage heaving and so detect when its breathing rate increases. But is that because it is afraid, or because it is responding in a stimulus-response–like manner to danger, gearing up for a fast output of energy in a flee response? Is it really, really afraid, or does it just share some of the characteristics of being afraid that we have come to understand in mammals? Perhaps evolution built in the stimulus-response mechanism because that enabled better survival rates. Perhaps only later with mammals did evolution add *real* fear.

Like mammals and reptiles, fish are vertebrates and share a similar body plan with steerable eyes. Rather than lungs, fish have gills, and we certainly get cues about their respiration state when we see them gasping on the ground behind a fisherman on a jetty. If we are snorkeling underwater, we are totally aware of their startle response and their ability to accelerate much faster than we could ever imagine doing. But when they come to rest again a few seconds later, it is hard to judge that they have been through any traumatic or stressful experience. They seem as calm as when we first encountered them. Or perhaps it is that they are just as fearful as when we first encountered them.

If we now look at insects and arachnids, it becomes much less clear that they are ever afraid; in fact, sometimes they seem quite fearless. Insects like cockroaches will flee from oncoming footfalls, but there are no indications whatsoever of fear. Their eyes are not steerable, and many have no discernible head motion relative to their bodies. And they do not bother to breathe—they simply respire through small tunnels inward from their body surfaces. Spiders and other arachnids sometimes have steerable eyes, but they are too small for us to see them. Like insects, they often flee from danger, but they do not show any other evidence of fear. On the other hand, social insects will often attack creatures many times their size; their genes have traded off the survival of the individual, who will not reproduce in any case, for the survival of the society, in which many of their genes will be reproduced.

By the time we get down to worms and arrive finally at amoebas, we are probably all willing to say that they do not have emotional responses. They are not something to which we can map our own fear responses. They are so utterly different and alien that they and we are incomparable at the level of our emotions.

These different sorts of animals appear to have different levels of emotional responses. It also happens that we as a society tend to treat them with different levels of ethical care.

Chimpanzees are usually viewed as very humanlike. There is great debate about the cruelty of keeping chimpanzees locked in cages, and indeed our zoological gardens have changed their policies on this over the last twenty years. Nowadays chimpanzees, as well as gorillas and orangutans, the other great apes that lie on the same twig of the evolutionary tree as humans, are normally only kept in large environments with indoor and outdoor spaces, and a freedom to roam within those expanded confines. The death of a single chimpanzee is treated very seriously, and a private citizen owning a chimpanzee would be roundly condemned and ostracized were he to arbitrarily have that animal killed, no matter how swiftly or humanely. He or she might well face criminal prosecution in many parts of the world.

We treat chimpanzees with an almost, but not quite, human level of respect. We do not give them the rights of humans, but we give them a very large measure of the respect that we give to humans. There are very few research experiments done with chimpanzees and almost none that involve invasive surgery. We as a society have raised the bar very high in how we demand that there must be clear and immediate payoff for humankind before we are willing to subject chimpanzees to the sorts of experiments that we are willing to allow macaque monkeys to routinely participate in. But Marc Hauser at Harvard points out that we do not quite give chimpanzees the level of "selfness" that we give ourselves. Imagine a race of intelligent beings showing up at our planet Earth and studying us for thirty years. Imagine how you personally would feel about having a team of these aliens assigned to you who would follow you everywhere you went for the next thirty years, day and night. They would follow you to work and sit off to the side observing. They would come along on your honeymoon and follow you into your bedroom. They would silently and unobtrusively follow you into the bathroom every time you went there, and would take a small sample. I doubt you would be very comfortable with this. We are, however, quite comfortable with some of the most admired animal researchers doing just this to chimpanzees out in their African habitats.

As we move to dogs and cats and horses, there are strict laws against cruelty to them. It is the right of their owners to have them killed for no particular reason, however, as long as it is done "humanely." This usually means quickly, and without pain or emotional upset. It is marginally more acceptable socially to have a cat put down than a dog or a horse— there is still something of a question of "Why?" for the latter cases. In all three cases, however, we treat these animals with respect.

Mice and rats get much less respect. We can buy traps at our local hardware store for killing these animals. When they work completely as designed, they break the necks of the animals, which die quickly. But it is not at all uncommon for the trap to break the spine of a mouse and for it to live on for some hours in pain. In our modern society that is certainly no one's business but that of the homeowner who is eliminating the pests. If, however, it became known that some particular homeowner was catching mice alive and then microwaving them to death, there would be outrage and criminal charges brought.

The level of respect we give to reptiles varies markedly from society to society. By the time we get to fish there is very little argument against letting a fish suffocate to death once it has been pulled from the sea or a stream. Some societies are happy to cook fish alive, while others are more squeamish about this practice.

Many of us spray and swat insects to death without a second thought.

We casually crush them with our shoes. We hang up high-temperature lamps that attract them and fry them to death. We inadvertently crush mites that live in our eyebrows and on our skin. We do not give these animals a first thought, let alone a second.

We grant animals some level of similarity to ourselves at an emotional level. That level of similarity happens to correspond fairly well to how similar they are to us in evolutionary and physiological ways.

Somewhere in that mixture of emotion and physiology we see enough similarity and have enough empathy that we treat animals that are similar to us in the moral ways that we have decided to treat other people.

Like Us

Is there, or will there ever be, enough similarity to us in our humanoid robots that we will decide to treat them in the same moral ways we treat other people and, in varying degrees, animals?

Our new humanoid robots have similar external form to us humans. That is, after all, why we call them humanoid. They have human shape and size and are able to move and respond in humanlike ways. To date, there are no robots, humanoid or otherwise, that have anything like a human physiology.

Our robots are all steel and silicon. Broadly, at least. They consist of many components made from plastics and metals. Some of them have an external skin, but they are not as soft and squishy as people. The actuators to date are all electric motors,[1] but the inclusion of springs and advanced control mechanisms have lead to very humanlike motions. Humanoid robot arms equipped this way are able to achieve subtle mo-

---

1. A few people have tried other actuators for humanoids with varying degrees of success, but all such actuators have less than the performance of electric motors. Disney displays use hydraulic motors for their animatronic humanoids that are set to repeat the same action, without variance, thousands of times per day. Others have used pneumatic actuators, but their control and performance is very poor, even at their best, and, the robots appear to have advanced Parkinson's disease.

tions in response to external forces. While still clumsy by human standards they are starting to make progress toward the sorts of interactions with objects that a four-or-five-year-old can do routinely. This is very different from the actions that we have associated with robots from seeing industrial robots pick up and place objects from precise locations and with incredible speed.

The source of energy for our humanoid robots is electricity from a wall socket. There is no imperative for them to gather, store, steward, or expend energy sparingly. At this point in time they do not need to engage in so many of the behaviors that we humans engage in, almost unconsciously, every day, to maintain our bodies and our existence. We must eat and drink every few hours to survive. We must sleep on a daily basis to remain healthy. We must breathe every few seconds or we die in a matter of minutes. As long as our humanoid robots are freely plugged into a wall socket, they have no need to do any of these things. There are a few early experiments that involve having robots gain their energy from digesting biological matter and using the resulting gases to run fuel cells. But as long as our robots are still primarily steel and silicon creations, that form of energy will be a novelty. It will always be possible to tap into an electricity source directly and bypass the physiological imperatives.

Sleep, however, is a different matter. The "reason" for sleep is still somewhat of a mystery. However, more and more recent studies show that at least part of the picture is that it is a time when short-term memories gathered during the waking period are consolidated into longer-term memory. It may turn out that there are legitimate reasons why our intelligent robots will have to shut down some of their interactive abilities in order to make such orderly compressions of recent experiences. If that is the case, then sleep may become something necessary to our humanoids in order for them to learn and adapt. On the other hand, it is plausible that we as engineers may be able to come up with something cleverer than evolution was able to do, and to generate algorithms that can do this consolidation while the robot is fully functional.

In the near term, the next ten to twenty years, say, it is a safe bet that our robots will remain very foreign in their physiology. It is likely that we will be able to make them more and more like humans in their external appearance. But we will always know that underneath their familiar exterior there is something very different from us.

So much for physiology. What about emotions? Are the emotions of robots at all like those of humans?

A number of robots that people have built, including Kismet and My Real Baby, are able to express emotions in humanlike ways. They use facial expressions, body posture, and prosody in their voices to express the state of their internal emotions. Their internal emotions are a complex interplay of many subsystems. Some have drives, such as Kismet's *loneliness* drive, that can be satiated only by particular experiences in the world, in this case detecting a human face. Some have particular digital variables within some of their many concurrent programs with explicit names like *arousal*. There are many interactions between the emotional systems, the perceptual systems, and the motor systems.

For instance, in Kismet, when the loneliness level is high, the visual system rates things that have skin color[2] as more interesting than the saturated colors of toys. As a result, there will be a higher tendency for Kismet to saccade toward things with skin color, which in its laboratory environment means that it will probably saccade to a human face if one is present. Human hands are the other skin-colored objects in its environment, but they are usually smaller and thus less attractive than faces. If the loneliness level rises, it will tend to shift Kismet's emotional state away from happiness. Kismet will more likely become fearful or angry in response to small annoyances from moving objects near its face. It will display those emotions through its facial expressions and vocalizations. As it gets unhappier and unhappier, lonelier and lonelier, it will refuse to look at a toy even if that is the only thing in view.

This is similar to what happens in the human brain, where there are primitive centers of emotion such as the amygdala and other parts of the limbic system. These structures receive inputs from many parts of the brain's perceptual subsystems, and at the same time innervate both primitive motor sections of the brain and the more modern decision-making and reasoning centers of the brain. Antonio Damasio, a neuro-

2. It is possible to detect the color of human skin across all races, for there is a fundamental visual characteristic that has just a little variation layered on top of it by the particular pigments produced by an individual. Whereas many people have tended to concentrate on differences in skin color, it is rather easy to build a robot visual system that recognizes the commonality.

scientist in Iowa, has explained the role of these structures in popular books about the relationship between emotions and more advanced centers in the brain. His message, in brief, is that emotions are both primitive in the sense that we carry around the emotional systems that evolution installed in our brains long before we had warm blood, and that they play intimate roles in all of the higher-level decisions that we tend to think of as rational and emotionless.

So our robots are being built with emotional systems that model aspects of what goes on inside the heads (and hearts) of humans. But is "model" the operative word here? Are they real emotions, or are they only simulated emotions? And even if they are only simulated emotions today, will the robots we build over the next few years come with *real* emotions? Will they need to be visceral emotions in the way that our dog can be viscerally afraid? What would it take for us to describe a response from a robot as visceral?

These are perhaps the deepest questions we, as humans, can ask ourselves about our technologies. The answers we choose to believe have significant impacts on our views of ourselves and our places in the universe. For if we accept that robots can have real emotions, we will be starting down the road to empathizing with them, and we will eventually promote them up the ladder of respect that we have constructed for animals. Ultimately we may have to worry about just what legal status certain robots will have in our society.

As we were building My Real Baby prototypes at iRobot Corporation, we built into them an emotional model. None of us were under any illusion that these were real emotions in the sense that a dog has emotions. The emotional system controls the behaviors in which the doll will engage and also triggers certain displays of emotion, so that the child knows what state the doll is in. We even called these displays *animations*: the series of facial expressions that the doll would go through, along with representative sounds (different in exact choice on every different occasion) that it makes. These were certainly fake emotions as far as we were all concerned.

Jim Lynch, a hard-nosed electrical engineer with a long career in building toys for other companies before ours, was the person responsible for designing the internal electronics for My Real Baby. As the prototypes got closer and closer to the version that was to be manufactured in bulk, we needed to do more and more testing. Within the company we set up a "baby-sitting" circle, including members of both the Toy Division and people who worked on other sorts of robots. For an hour at a time the prototype robot baby doll would be in the charge of a different person. They were to note what went wrong with it, if anything, and to try to record what had happened right before a failure. In this way we were able to detect bugs deep within the complex interactions of the hundreds of concurrent programs that generated the doll's overall behavior, and the complex electronics by which that software ran, sensed, and acted.

One day Jim had just received a doll back from a baby-sitter. As it lay on the desk in his office, it started to ask for its bottle: "I want baba." It got more and more insistent as its hunger level went up, and soon started to cry. Jim looked for a bottle in his office but could not see one. He went out to the common area of the Toy Division and asked if anyone had a bottle. His doll needed one. As he found a bottle and rushed back to his office to feed the baby, a realization came over him. This toy, that he had been working on for months, was different from all previous toys he had worked on. He could have ignored the doll when it started crying, or just switched it off. Instead, he had found himself *responding* to its emotions, and he had changed his behavior as though the doll had had real emotions. He was extraordinarily happy about what he had helped create. But here we see the crux of the question. Can a $100 doll have emotions, or can it merely trigger the responses that we have built into us evolutionarily, that over the eons have been honed to respond to emotions in other *real* creatures?

It is all very well for a robot to *simulate* having emotions. And it is fairly easy to accept that the people building the robots have included *models* of emotions. And it seems that some of today's robots and toys *appear* to have emotions. However, I think most people would say that our robots do not really have emotions. As in the cases of Sherry Turkle and Jim Lynch, our robots can possess us for a moment, making us act as if they have emotions and are beings, but we do not stay fooled for long. We treat our robots more like dust mites than like dogs or people. For us the emotions are not real.

Birds can fly. Airplanes can fly. Airplanes do not fly exactly as birds do, but when we look at them from the point of view of fluid mechanics, there are common underlying physical mechanisms that both utilize. We are not tempted to say that airplanes only simulate flying. We are all happy to say that they really do fly, although we are careful to remember that they are not flapping their wings or burning sugar in their muscles—they do not even have muscles. But they do fly.

Is our question about our robots having real emotions rather than just simulating having emotions the same sort of question as to whether both animals and airplanes fly? Or is there something deeper in the semantics of the word "emotion" as contrasted to "flying" that makes the former impossible for a machine to have but not the latter?

Specialness

Over the last four thousand years mankind has undergone some very large changes in worldview. Perhaps the issue of whether robots can have emotions, even in principle, let alone now, will be the topic of another change in worldview.

Human thought and intellectual discourse over the last five hundred years has been a history of changing understanding of mankind's place and role in the universe. Early on, the existence of a god, or gods, was postulated to explain the inexplicable. This mechanism has been very common across all human cultures. The details of the form of the gods have varied. There have been all-powerful single deities, and panoplies of flawed humanlike actors, and spirits that are at the edge of our understanding. But all these gods have had wills of their own, and have done things for their own reason, which is beyond our reason. We have been saved the need to explain everything in the universe. It has been beyond our place in the universe to understand or question the details. Religions have provided comfort where faith and a limit on questioning replaces the terrible unknown, the gnawing lack of understanding. Mysticism is comforting to us in the way it releases us from having to search further. It provides a satisfaction and an ease. It utilizes those early parts of the brain that are devoted to emotions and sends signals of contentment throughout our brain making us feel good.

In case you are explicitly religious and feel that your beliefs are being attacked by these remarks, you might perhaps reflect that they need only apply to all those other people throughout history who have worshiped other gods. Their beliefs and yours are not consistent, and so they must ultimately be mistaken. My remarks need only apply to why they are so committed to their incorrect beliefs.

And in case you are an atheist, let me remind you that you most probably have similar deep-seated beliefs, which while not openly religious share many of the same dogmatic aspects of religion. These beliefs might include faith in rationalism, science, bionomics, extropianism, or technofuturism. All of these also trigger the same sorts of satisfaction in that they provide a belief system that does not need to be questioned. They provide a place of contentment for our brains where we do not need to desperately search for answers, where we are not triggering the same responses that enabled us to think quickly and survive dangers on the savanna.

My point here is that all of us, myself included, have our own sets of deep-seated beliefs that we prefer to keep as beliefs. Our usual reaction to having them challenged is anger—a primitive fight response. Our emotional systems kick in and provoke us to try to eliminate the danger through the quickest means possible. We generally do not savor the feelings of internal conflict that thinking about our deepest-held beliefs brings upon us.

Over the last five hundred years, however, mankind as a species has been confronted again and again with challenges to our most deeply held beliefs. We have had to go through multistage processes at each of these challenges and find ways to reconcile new understandings of the universe with what we have already believed. These confrontations have always been emotional, and often they have been violent. They have taken tens to hundreds of years to become accepted as general truths, to become assimilated into the belief systems that we all hold and that provide us with the comfort levels that let us reach a state of equilibrium in our brain chemistry.

The natural and prescientific view that human cultures all seem to have adopted is that the Earth is the center of the universe, with the Sun, Moon, and stars revolving around it. As scientific observations were taken, this view was challenged. In the sixteenth century Copernicus suggested a heliocentric universe, with the Earth revolving around it.

Later that century Tycho Brahe, with his reams of meticulous data, worked on showing that the planets circled the Sun but was not ready to give up the Earth as the center of the universe. Kepler, early in the seventeenth century, armed with Brahe's data, was able to break out of the belief in the simplicity of circles and show that the planets moved in elliptical orbits. The heliocentric universe by this time was an idea that was in the air, and discussed, but it came to a head in just a few years with Galileo. He was the first to turn a telescope to the sky, early in the seventeenth century, and before long the essential truth of the Sun as the center of our solar system become evident to him. This was not acceptable to the Catholic Church, and Galileo prudently chose the path of repentance rather than sticking with his truth.

The church could not accept that the terra firma on which mankind stood could not be the center of the universe. It had to be. For otherwise, why had God placed us somewhere peripheral? How could God, who had created men, for whatever reason, as the embodied masters of the universe, not place them centrally? Mankind was special. Specialness demanded a central place in the universe. Heaven was above. Hell below. The center of action was the Earth. That was mankind's place in the universe and as such it must be the center.

To accept that the Sun was the center rather than the Earth demanded some acceptance of the lack of specialness of humans. As the idea swirled around for decades and more and more observations confirmed it, the idea was finally accepted and became mainstream. But it required accommodation in theologies, and it was the beginning of a dual belief system that many adopted. Religion taught about aspects of humanhood, while science gave the details of the universe. Mankind became separated from the physical world in some ways. As before, mankind was still special in nonphysical ways. Mankind was not subject to the same sorts of questioning and investigation as the physical world. Man's body was sacred and not subject to posthumous dissection.

Over time the church changed, and before long Jesuit priests had become master observational astronomers, chronicling the motions of the universe. Over the next few hundred years they were active in further reduction of the specialness of our home—Earth—within the universe. With the aid of telescopes it became clear that our sun was just one of many, and ultimately we learned that our sun was not at the center of those many, but off to one side in a galaxy of billions of suns. Later we realized that our galaxy was but one of billions of galaxies. The universe is vast, and we and everything we experience directly are just one tiny speck of a vast existence.

Mankind's place in the universe became distinctly nonspecial. But mankind's place on Earth was still very special. Man was the lord of his dominion, the dominion of Earth. Animals were very different from people. God had put them both on the Earth, but it was only humans that worshiped God, only humans with a special relationship with God. Furthermore, at least for Christians, God had created man in his own image. Whether that was the case, or whether just the opposite, that man had created God in his own image, need not concern us for now. Mankind had a special place in the universe. No longer the physical center of the universe, but still the center in God's view of the universe, for man had been created differently from all other creations.

This special place was shaken mightily by the ideas of evolution that culminated in Charles Darwin's book *On the Origin of Species*. This was cataclysmic for people's view of themselves as being special. Not only were they closely related to apes, they were descended from the very same forebears as were the monkeys in the zoos. Monkeys and people were cousins. Images of this relationship were used in cartoons to ridicule Darwin—how could it possibly be said that humans and monkeys shared blood? Not only were all the "inferior" races of man closely related to educated people of the day, but even monkeys were kin. This loss of specialness was hard to bear.

The shock that it brought was too much for many, and even today, supposedly educated people in the United States cling to a belief that it is simply wrong. Despite the preponderance of evidence from the fossil record to experiments that any high school student can carry out in a well-equipped laboratory, people cling to the notion that it is some wrongheaded theory of scientists. The number of facts that need to be explained away by some alternative theory is simply too large. There is no rational explanation for what we see about us other than the truth of Darwin's central ideas. The details here and there may be fuzzy, but the central outline is undeniable. The last remaining bands of true disbelievers from the intellectual badlands of the United States plead for keeping our children in ignorance, as this is the only tool left to them in their battle against the tidal wave of evidence. Why do they carry out this battle? Why do they cling to their flat-Earth–like beliefs? They are afraid of losing their specialness. For those whose faith is based on literal interpretations of documents written by mystics thousands of years ago when just a few million people had ever lived, the facts discovered and accumulated over thousands of years by societies of billions of people are hard to accommodate.

Much of Western religion was able to accommodate evolution. It is hard to say whether it was the accommodators or the rejecters who were more intellectually consistent. In any case, a new model soon arose. God had created the clockwork of evolution and this was the mechanism, along with his ever present guiding hand, which had enabled the miracle that is us to evolve. There had been an *upward* movement throughout evolution toward us as the pinnacle. By the late nineteenth century there were illustrations of monkeys evolving from all fours to upright walking, through the chimpanzee and gorilla, through the "lesser races" of humans from Africa, to the top, to the special, to the white European male.

This view let people retain some specialness for themselves, but it was dead wrong when the true tenets of evolution were examined carefully. In evolution there was no special place or destination for mankind. It was all rather random, with luck playing out at every step of the way. A slightly delayed cosmic ray at some particular time might have led to an entirely different set of species on the surface of the Earth today, with mankind nowhere to be seen. This was the place for God's hand to ensure that all worked out for the best in the long run, ensuring that our special species would be here. This was how to accommodate evolution, but retain specialness.

Throughout the twentieth century there were many more blows to mankind's specialness, though none as severe as the loss of the Earth-centric universe or the loss of the direct creation of mankind by God. Perhaps in reality some of these blows should be seen as severe, but the vast majority of people have still not taken them seriously nor thought through their consequences.

To a society that was getting used to the idea that science unveiled secrets of the universe and led to an orderly explanation of all that happened, physics provided three unwelcome blows. These were the theory of relativity, quantum mechanics, and the Heisenberg uncertainty principle. Each of these in its own way weakened the idea that it was possible for a human to ultimately know what was going on in the universe. People's senses were not the ultimate arbiter of what was what in the universe. There were things that were ultimately not able to be sensed. These three realizations of physics ate away even more at the idea of the specialness of man—man could not know ultimate truths because they were unknowable.

When Crick and Watson discovered the structure of DNA, there was a further weakening of specialness. Soon it was found that all living creatures on Earth shared the same sort of DNA molecules and used the same coding scheme to transcribe sequences of base pairs into amino-acid building blocks of proteins. The proteins so produced were the mechanism of the cell that determined whether it was a bacterium, the skin of a lizard, or a human neuron in prefrontal cortex. But it was all the same language. Even worse, we humans shared regulatory genes, relatively unchanged, with animals as simple as flies. There were even clear relationships between some genes in humans and in yeast. In early 2001 the two human genome projects announced that there were probably only about 35,000 genes in a human—disappointingly few—less than twice as many as in a fruit fly. These discoveries hammer home the nonspecialness of humans. We are made of the same material as all other living things on this planet. Furthermore, our evolutionary history and relationship to all other living things is clear to see. We share some 98 percent of our genome with chimpanzees. We really are just like those animals, separated by a sliver of time in evolutionary history.

Machines

The two major blows to our specialness, that Earth is not the center of the universe and that we evolved as animals, did not come suddenly upon us. In each case, there were precursors to Galileo and Darwin, debates and vituperative arguments. These lasted for decades or centuries before the crystallizing events and for decades or centuries afterward. Indeed, it is only with the hindsight of history that we can point to Galileo's trial and Darwin's publication as the crystallizing events. They may not have seemed so out of the continuum to the participants or observers.

We have been in the beginning skirmishes of the third major challenge to our specialness for a period of almost fifty years now. We humans are being challenged by machines. Are we more than machine, or can all of our mental abilities, perceptions, intuition, emotions, and even spirituality be emulated, equaled, or even surpassed, by machines?

The machines we had been building for the last few millennia had never really challenged our specialness until recently. They were sometimes stronger or faster than us, or could fly while we could only walk and run, but animals could do all those things too.

With the invention of the computer and some clever programming of them in the 1950s and '60s, machines started to challenge us on our home turf: syntax and technology. As artificial intelligence proceeded, machines regularly started operating in arenas that had previously been the domain of unique human capabilities.

Early programs enabled computers to prove theorems in mathematics—not very hard ones at first, but proofs from the first few pages of the book *Principia Mathematica* by Bertrand Russell and Alfred North Whitehead, an exceedingly erudite attempt at formalizing mathematics. Computers were doing something with an intellectual flavor that only the most educated people could understand. There was also early progress on computers processing human languages, although there was a long hiccup from the mid-sixties to the mid-nineties without much outwardly visible progress. But those early programs, along with Noam Chomsky's generative grammars for all human languages, showed that perhaps language and syntax could be explained through fairly simple sets of rules, implementable on computers. By 1963 there was a computer program that could do as well as MIT undergraduates on the MIT calculus exam. By the mid-sixties, computers were playing chess well enough to compete in amateur chess tournaments.

As Hollywood caught up with what was happening in the research laboratories, there started to be movies about electronic brains, and the intelligence of machines perhaps rivaling that of humans. The questions of whether this was possible in principle, and whether it was likely technologically, gained credence. While Hollywood predicted the overwhelming of mankind by machines, a serious debate started in academia.

Those who took the strong AI position that of course machines would soon be superior to humans felt it was ridiculous that such a claim could be questioned. A number of people took up the challenge to defend the honor of humans. Many of them had good reasons to question the state of the art in computer intelligence, but they were often hoisted on their own petards by making weak counterattacks that were easy to foil.

When I was young and first read the children's book *Robots and Electronic Brains,* I noticed that although it was very enthusiastic about the technology it also had one little section showing that all was not marvelous in the world of computers. It pointed out that a few years earlier (the book was published in 1963) a "Chinese-American bookkeeper with an abacus won a race against an electronic calculating machine." Although the machines were blazingly fast, a human could still keep up! All was not lost for human intellect. As computers get roughly twice as fast every two years, any such advantage for a human, in terms of pure speed, is a losing proposition. Our computers eventually pass any fixed speed mark and are soon unimaginably faster.

Alan Turing, the British mathematician who had formalized much of computation in the thirties, wrote up a thought experiment in a 1950 paper. He started out by supposing that there was a person at the other end of a teletype (think of it as a two-person chat room with a very noisy keyboard), and you had to determine whether they were male or female by asking them questions. There was no requirement that they be truthful in answering. Then instead consider the case that you do not know whether it is a person or a computer at the other end of your link. Turing claimed that if a computer could fool people into thinking it was a person, then that would be a good test of intelligence. This has come to be known as the Turing test and has been the source of much debate, as well as a few interesting demonstrations.

Joseph Weizenbaum at MIT wrote a program in 1963 called ELIZA, which played the role of a psychiatrist. It carried out simple syntactic conversions on sentences typed into it, and did a few simple word substitutions resulting in questions that resembled those that a TV version of a Rogerian psychiatrist would ask. Although the program was simple, and clearly it did not understand language very well, some people would spend hours pouring out their hearts to it. Weizenbaum used this as evidence that the Turing test was flawed. He was distressed that people would share their intimate thoughts with the program and used its existence to argue that artificial intelligence was doomed. He had been shocked that people took his program seriously and that practicing psychiatrists, and others, speculated on the day that real psychotherapy might be provided by a machine. Worse, he was appalled at their own self-image, when they compared what might be doable by a computer to their own professional activities.

> What can the psychiatrist's image of his patient be when he sees himself, as therapist, not as an engaged human being acting as a healer, but as an information processor, following rules, etc.?

Weizenbaum was fleeing from the notion that humans were no more than machines. He could not bear the thought. So he denied its possibility outright.

Jaron Lanier, on John Brockman's www.edge.org Web site, in the year 2000 made a similarly confused claim about the futility of trying to build intelligent computers. People listen to Jaron because he is one of the founders of virtual reality and a technical wizard. He set out an argument that the Turing test is flawed because there are two ways that it can be passed. One is that computers can get smarter, or the second is that people can get dumber. He goes on to give examples of how people are slaves to machines with bad software, including those that assess our credit ratings. This, he claims, is people getting dumber to accommodate bad software—a claim that may well be valid. But then he states that this shows that "there is no epistemological difference between artificial intelligence and the acceptance of badly designed computer software."

And with that he dismisses work in artificial intelligence as being based on an intellectual mistake. Unfortunately for Jaron his conclusion does not follow from his premise. He has neither refuted the Turing test as a valid test for machine intelligence (I happen to be fairly neutral about whether it is a good test or not) nor shown that the failure of that test means that it is impossible to build intelligent machines.

Weizenbaum and Lanier, both respected for their work in fields other than intelligent machines, wade into an argument and make nonsensical statements. Their aim is to discredit the idea that machines can be intelligent. My reading of them both is that they are afraid to give up on the specialness of mankind. An intelligent machine will call into question one of humankind's last bastions of specialness, and so they deny, without rational argument, that such a machine could exist. It is intellectually too scary.

Hubert Dreyfus, a Berkeley philosopher, has fought a more explicit rearguard action against intelligent machines for over thirty-five years. In 1972 he published a book that set out what computers were incapable of doing. I think he was right about many issues, such as the way in which people operate in the world is intimately coupled to the existence of their body in that world. Unfortunately he made claims about what machines could not do in principle. He made a distinction between the nonformal aspects of human intelligence and the very formal rules that make up a computer program. Many of his arguments make such distinctions, and he explicitly stated that the nonformal aspects, the judgments, recognition of context, and subtle perceptual distinctions could not be reduced to mechanized processes. He was not able to supply any evidence for such a claim other than rightly pointing out that, at the time, AI researchers had not been able to produce that sort of informal intelligence. His mistake was to claim that these arguments were *in principle* arguments against any ultimate success. Many of the things that he said were not possible in principle have now been demonstrated.

Dreyfus was well known to the artificial intelligence community by the mid-sixties, as he had made some rather public statements about the inability of computers to play chess. He rightly pointed out that humans did not play chess the way machines were programmed to play chess in the 1960s. The machines followed the original ideas of Alan Turing, Norbert Wiener, and Claude Shannon, who had all suggested searching a tree of possible moves. Here Dreyfus's mistake was to get befuddled by the algorithmic nature of this procedure. To him it was impossible that simply following the rules of exhaustively trying moves, down to some fixed number of moves looking ahead, and evaluating the resulting positions, could generate good play. People were much more organic. They had insight. A rule-based system to play chess could not have insight.

Rather rashly Dreyfus, himself a rather poor amateur chess player, claimed that no machine would ever beat him at chess. Richard Greenblatt at the MIT Artificial Intelligence Laboratory was happy to oblige him, and his MacHack program beat Dreyfus in the first and only encounter in 1967.[3] Unbowed, Dreyfus declared that a program would never beat a good chess player, one who was nationally ranked. When that milestone was reached, Dreyfus claimed that a computer would never beat the world champion. As we all know, this did happen in 1997 when IBM's Deep Blue program beat world champion Garry Kasparov.

Dreyfus was dead right that the computer did not play chess at all like a human. But he was fooled by his own unease that intelligence could emerge from things that were described algorithmically. We saw in chapter 3 how lifelike animal behavior can arise from a complex set of interactions across a set of simple rules. Dreyfus did not understand that something could emerge from simple rules applied relentlessly that was able to do better at a task than an extremely gifted human.

We see now a pattern starting to emerge. Very bright people have drawn lines in the sand, saying that computers will not be able to achieve some sort of parity with humans. Again and again, they have had to erase those lines and draw new ones. This has happened often enough that I think a consensus has arisen that computers are able not only to calculate better than humans but to do many other tasks better than humans.

Computers are now better at doing symbolic algebra than are humans. During the sixties and seventies scientists and engineers got into the habit of using their institution's central computer to do large-scale numerical calculations. By the eighties they were even doing small-scale calculations on machines—a desktop or hand calculator. They still had to do the algebra and calculus by themselves, determining what formulas that were to be used for the calculation. That required too much in-

3. Richard Greenblatt was a colorful figure at the MIT Artificial Intelligence Laboratory who built the first successful chess-playing programs. All the ideas that were in Deep Blue when it beat Garry Kasparov in 1997 were present in Greenblatt's program in the mid-sixties. All that had changed was that more brute force computing power was available so that Deep Blue could evaluate 200 million positions per second compared to the few hundred that Greenblatt's program could handle.

sight to be done by machine. Researchers such as Jim Slagle and Joel Moses at the MIT Artificial Intelligence Laboratory worked through the sixties on programs that could manipulate symbolic mathematics, search through possibilities, and apply heuristics. Now most scientists and engineers use programs like Mathematica or MATLAB to do their symbolic algebra and calculus for them. The machines are better at it than people are, even with their insight.

Computers are also better at designing certain classes of numerical computer programs (filters for digital signal processors) than are humans. They are better at optimizing large sets of constraints, such as simultaneously setting prices for many interlocked goods and services. They are better at laying out circuits for fabrication. They are better at scheduling complex tasks or sequences of events. They are better at checking complex computer protocols for bugs. In short, they are better at many things that were previously the purview of highly trained mathematically inclined human experts. Those experts, when carrying out their work would in the past have been described as *thinking,* or *reasoning,* as they worked on their design or optimization problems.

While we have surrendered our superiority in these abilities, these formerly uniquely human abilities, to machines, we have not given up on everything. Most people would agree today that our computers and their software are not usually able to have true deep insights nor are they able to reason across very different domains, as we are apt to do.

Just as we are perfectly willing to say that an airplane can fly, most people today, including artificial intelligence researchers, are willing to say that computers, given the right set of software and the right problem domain, *can* reason about facts, *can* make decisions, and *can* have goals. People are also willing to say it is possible to build robots today that *act as if* they are afraid, or that *seem* to be afraid, or that *simulate* being afraid. However, it is much harder to find someone who will say that today's computers and robots can be *viscerally afraid.*

We draw a line at our emotions. In fact, we use derisive language about machines lacking emotions. We talk about "cold, hard machines" to indicate that they have no emotional component. Despite Kasparov's perception of Deep Blue as having made insightful plans, he was rather gleeful that despite its win it did not enjoy winning or gain any satisfaction from it.

We may have lost our central location in the universe, we may have lost our unique creation heritage different from animals, we may have been beaten out by machines in pure calculating and reason, but we still have our emotions. This is what makes us special. Machines do not have them, and we alone do. Emotions are our current last bastion of specialness. Interestingly we allow some emotions to animals—we increase our tribe to include them in the exclusion of machines from our intimate circle. Within the animal empire we are still at ease about our superior place, most of us having adapted to being related to our furry companions in our post-Darwin world.

Further Reading

Damasio, A. R. 1999. *The Feeling of What Happens: Body and Emotion in the Making of Consciousness*. New York: Harcourt Brace Jovanovich.

Dreyfus, H. L. 1972. *What Computers Can't Do*. New York: Harper & Row.

Dyson, G. B. 1997. *Darwin Among the Machines*. Reading, Mass.: Addison-Wesley.

Greenblatt, R., D. Eastlake, and S. Crocker. 1967. "The Greenblatt Chess Program." In *Proceedings of the Fall Joint Computer Conference*, pp. 801–10. Montvale, N. J.: AFIPS Press.

Turing, A. M. 1950. "Computing Machinery and Intelligence." *Mind* 59: 433–60. The article also appeared in G. F. Luger, ed. 1995. *Computation and Intelligence: Collected Readings*. Cambridge, Mass.: MIT Press.

Turkle, S. 1995. *Life on the Screen*. New York: Simon & Schuster.

Weizenbaum, J. 1976. *Computer Power and Human Reason*. New York: Freeman.

We Are Not Special

I f we accept evolution as the mechanism that gave rise to us, we understand that we are nothing more than a highly ordered collection of biomolecules. Molecular biology has made fantastic strides over the last fifty years, and its goal is to explain all the peculiarities and details of life in terms of molecular interactions. A central tenet of molecular biology is that *that is all there is*. There is an implicit rejection of mind-body dualism, and instead an implicit acceptance of the notion that mind is a product of the operation of the brain, itself made entirely of biomolecules. We will look at these statements in more detail later in the chapter. So every living thing we see around us is made up of molecules—biomolecules plus simpler molecules like water.

All the stuff in people, plants, and animals comes from transcription of DNA into proteins, which then interact with each other to produce structure and other compounds. Food, drink, and breath are also taken into the bodies of these organisms, and some of that may get directly incorporated into the organism as plaque or other detritus. The rest of it either reacts directly with the organism's biomolecules and is broken down into standard components—simple biomolecules or elements that bind to existing biomolecules—or is rejected and excreted. Thus almost everything in the body is biomolecules.

Biomolecules interact with each other according to well-defined laws. As they come together in any particular orientation, there are electrostatic forces at play that cause them to deform and physically interact. Chemical processes may be initiated that cause one or both of the molecules to cleave in interesting ways. With the totality of molecules, even in a single cell, there are chances for hundreds or thousands of different intermolecular reactions. It is impossible then to know or predict exactly which molecules will react with which other ones, but a statistical model of the likelihood of each type of reaction can be constructed. From that we can say whether a cell will grow, or whether it provides the function of a neuron, or whatever.

The body, this mass of biomolecules, is a machine that acts according to a set of specifiable rules. At a higher level the subsystems of the machine can be described in mechanical terms also. For instance, the liver takes in certain products, breaks them down, and recycles them. The details of how it operates can be divined from the particular bioreactions that go on within it, but only a few of those reactions are important to the liver itself. The vast majority of them are the normal housekeeping reactions that are in almost every cell in the body.

The body consists of components that interact according to well-defined (though not all known to us humans) rules that ultimately derive from physics and chemistry. The body is a machine, with perhaps billions of billions of parts, parts that are well ordered in the way they operate and interact. We are machines, as are our spouses, our children, and our dogs.

Needless to say, many people bristle at the use of the word "machine." They will accept some description of themselves as collections of components that are governed by rules of interaction, and with no component beyond what can be understood with mathematics, physics, and chemistry. But that to me is the essence of what a machine is, and I have chosen to use that word to perhaps brutalize the reader a little. I want to bring home the fact that I am saying we are nothing more than the sort of machine we saw in chapter 3, where I set out simple sets of rules that can be combined to provide the complex behavior of a walking robot. The particular material of which we are made may be different. Our physiology may be vastly different, but at heart I am saying we are much like the robot Genghis, although somewhat more complex in quantity but not in quality. This is the key loss of specialness with which I claim mankind is currently faced.

And why the bristling at the word "machine"? Again, it is the deep-seated desire to be special. To be more than mere. The idea that we are machines seems to make us have no free will, no spark, no life. But people seeing robots like Genghis and Kismet do not think of them as clockwork automatons. They interact in the world in ways that are remarkably similar to the ways in which animals and people interact. To an observer they certainly seem to have wills of their own.

When I was younger, I was perplexed by people who were both religious and scientists. I simply could not see how it was possible to keep both sets of beliefs intact. They were inconsistent, and so it seemed to me that scientific objectivity demanded a rejection of religious beliefs. It was only later in life, after I had children, that I realized that I too operated in a dual nature as I went about my business in the world.

On the one hand, I believe myself and my children all to be mere machines. Automatons at large in the universe. Every person I meet is also a machine—a big bag of skin full of biomolecules interacting according to describable and knowable rules. When I look at my children, I can, when I force myself, understand them in this way. I can see that they are machines interacting with the world.

But this is not how I treat them. I treat them in a very special way, and I interact with them on an entirely different level. They have my unconditional love, the furthest one might be able to get from rational analysis. Like a religious scientist, I maintain two sets of inconsistent beliefs and act on each of them in different circumstances.

It is this transcendence between belief systems that I think will be what enables mankind to ultimately accept robots as emotional machines, and thereafter start to empathize with them and attribute free will, respect, and ultimately rights to them. Remarkably, to me at least, my argument has turned almost full circle on itself. I am saying that we must become less rational about machines in order to get past a logical hangup that we have with admitting their similarity to ourselves. Indeed, what I am really saying is that we, all of us, overanthropomorphize humans, who are after all mere machines. When our robots improve enough, beyond their current limitations, and when we humans look at them with the same lack of prejudice that we credit humans, then too we will break our mental barrier, our need, our desire, to retain tribal specialness, differentiating ourselves from them. Such leaps of faith have been necessary to overcome racism and gender discrimination. The same sort of leap will be necessary to overcome our distrust of robots.

Resistance Is Futile

If indeed we are mere machines, then we have instances of machines that we all have empathy for, that we treat with respect, that we believe have emotions, that we believe even are conscious. That instance is us. So then the mere fact of being a machine does not disqualify an entity from having emotions. If we really are machines, then in principle we could build another machine out of matter that was identical to some existing person, and it too would have emotions and surely be conscious.

Now the question is how different can we make our Doppelgänger from the original person it was modeled upon. Surely it does not have to be precisely like some existing person to be a thinking, feeling creature. Every day new humans are born that are not identical to any previous human, and yet they grow to be a unique emotional thinking, feeling creature. So it seems that we should be able to change our manufactured human a little bit and still have something we would all be willing to consider a human. Once we admit to that, we can change things some more, and some more, and perhaps eventually build something out of silicon and steel that is still functionally the same as a human, and thus would be accepted as a human. Or at least accepted as having emotions.

Some would argue that if it was made of steel and silicon, then as long as people did not know that, they might accept it as human. As soon as the secret was out, however, it would no longer be accepted. But that lack of acceptance cannot be on the basis that it is a machine, as we are already supposing that we are machines. Indeed, the many arguments that abound about why a machine can never have real emotions, or really be intelligent, all boil down to a denial of one form or another, that we are machines, or at least machines in the conventional sense.

So here is the crux of the matter. I am arguing that we are machines and we have emotions, so in principle it is possible for machines to have emotions as we have examples of them (us) that do. Being a machine does not disqualify something from having emotions. And, by straightforward extension, does not prevent it from being conscious. This is an assault on our specialness, and many people argue that it cannot be so, and argue that they need to make the case that we are more than machines.

Sometimes they are arguing that we are more than conventional computers. This may well be the case, and I have not, so far, taken any position on this. But I *have* taken a position that we are machines.

Many hard-nosed scientists have joined this intellectual fray. Some do better than others in making their arguments. But usually those that argue that machines will never be as good as us somehow end up arguing that we are more than mere machines. Often the players are strong materialists who proclaim that they are not arguing for any role from God, or a spirit, or a soul for people, and not even some élan vital, or life force. Rather they argue that there is something explicitly or implicitly more than a mere machine but which is at the same time of the material world. In order to make such an argument they need to invent some sort of *new stuff*. Often they deny that this is what they are doing. Below, I pick Roger Penrose, David Chalmers, and John Searle as representatives of the people who provide the most cogent arguments against people being just machines. Penrose and Chalmers could turn out to be right, although neither of them provides any data to support his hypothesis. Searle, I think, is just plain confused.

Roger Penrose, the British physicist and mathematician, is certainly a hard-nosed scientist. He has written a number of books that attack artificial intelligence research as not being likely to build intelligent, conscious machines. While he might be right about that, his arguments are flawed and are certainly good examples of a "new stuff" argument.

Penrose starts out by making a mistake in understanding Gödel's theorem and Turing machines. Kurt Gödel shocked the world in the thirties by showing that any consistent set of axioms for mathematics had theorems that could not be proved within that set of axioms. Turing machines, the formalization of modern computers that Alan Turing came up with around the same time, operate within whatever set of axioms they are given. So combined, these two results say that there are true theorems within mathematics that a computer cannot prove to be true. So if human mathematicians were computers, there would be theorems that they could not prove. This rankles Penrose's pride in himself and his friends. They are all able to prove lots of theorems, so Penrose incorrectly concludes that humans, or at least mathematicians, cannot be computers.

Now Penrose is in a fix. He is a hard-core materialist, but the material world cannot explain what he thinks he has observed. He needs to find something new to add within the material world that is outside the realm of ordinary computers. He hits upon the microtubules inside cells where there are quantum effects in operation. He hypothesizes, with no data to support his hypothesis, that the quantum effects there are the source of consciousness. Rather than accept the idea that consciousness is just an extension of the ideas we described in chapter 3, the result of simple mindless activities coupled together, Penrose asserts that there is some inexplicable extra thing, quantum mechanics, at play in living systems.

Penrose, in his bid for scientific materialism, has resorted to a mysterious higher force. Rather than accept the offensive idea that his magnificent mind is the product of simple mechanisms playing out together, he calls into play something too complicated for us to understand fully. He invents his own little deity, the god of quantum mechanics.

David Chalmers, a philosopher at the University of Santa Cruz, is another hard-nosed materialist. He invents a very different type of "new stuff." In his version there may be some fundamentally new type in the universe that we have not yet observed directly. He compares it to *spin* or *charm* in particle physics—properties of subatomic particles. Neither of these can be reduced to mass or charge, or any of the more usual types we have come to understand from classical physics. In his theory there is perhaps another type like these that we cannot observe directly with our senses, just as we cannot observe spin or charm. This new type may then be the basis of consciousness. Again, we see an appeal to some sort of higher authority, something different from anything that is rule-based, or mechanismlike. Chalmers appeals to something mysterious and not understood so that he can save his materialist view of the world without having to give up his specialness, to be nothing more than a mere machine.

John Searle is a very well respected philosopher at Berkeley. He professes to believe that mind and particularly consciousness is an emergent property of our having a brain inside our skulls. He claims to be interested in scientific explanations. But deep down he reveals that he is completely unable to accept that anything but real neurons can produce consciousness and understanding. He occasionally admits as much, for example:

> In this case, we are imagining that the silicon chips have the power . . . to duplicate the mental phenomena, conscious and otherwise. . . .
> I hasten to add that I don't for a moment think that such a thing is even remotely empirically possible. I think it is empirically absurd to suppose that we could duplicate the causal powers of neurons entirely in silicon.

Beyond this his arguments get circular, although I am sure he would rather pejoratively disagree with my analysis here.[1] His arguments are largely of the form of imagining robots and computer programs that have the same input-output behaviors as animals or people, and then claiming that the (imagined) existence of such things proves that mental

---

1. When Kasparov was beaten by Deep Blue, Searle said it was a "hunk of junk, designed by someone," and meant nothing. In some ways I agreed with him, but I was so jealous. I wished he had insulted my robots like that. I would have taken it as a badge of honor.

phenomena and consciousness exist as a property of the human brain, because the robot has neither of them, even though it is able to operate in the same way as a human. This argument as I have related it may not make sense to you, but I think it is a fair summary of Searle's position. And I think the real basis for his arguments, deep down, emotionally, is that he does not want to give up the specialness of being human.

His most famous argument concerns his "Chinese Room." John Searle does not understand Chinese. He talks about the equivalent of a computer program, a set of instructions that takes inputs in Chinese and outputs answers in Chinese. Some such computer programs exist today, as long as the domain of discussion remains somewhat limited. Searle argues that if he were locked in a room with these instructions, written out in English, and a slip of paper with a question written in Chinese was passed under the door, he could follow all the rules, written up in a large book for him, and eventually pass back under the door the answer written in Chinese on a slip of paper.

Searle correctly states that he, John Searle, would still not understand Chinese. He then absurdly concludes that no computer can understand Chinese. But he is making a fundamental mistake. Just as no single neuron in a Chinese speaker understands Chinese, Searle, a component of a larger system, does not need to understand Chinese for the whole system to understand Chinese. Searle goes on to deny that the whole system, the room, and he are conscious. I claim that we just do not know whether that is possible or not. In principle, I certainly think it is possible, but perhaps there would need to be some mechanism beyond just a set of rules for Searle to interpret.

Of course, as with many thought experiments, the Chinese Room is ludicrous in practice. There would be such a large set of rules, and so many of them would need to be followed in detailed order that Searle would need to spend many tens of years slavishly following the rules and jotting down notes on an enormous supply of paper. The system—Searle and the rules—would run as a program so slowly that it could not be engaged in any normal sorts of perceptual activity. At that point it does get hard to effectively believe that the system understands Chinese by any usual understanding of "understand." But precisely because it is such a ludicrous example, slowed down by factors of billions, any conclusions from that inadequacy cannot be carried over to making conclusions about whether a computer running the same program "understands" Chinese.

My conclusion after reading Searle's arguments is that he is afraid to give a machine consciousness. He claims it is something special about human brains and neurons in particular, but he never gives a hint of what it is that makes them special. Apart from the fact that they give rise to consciousness. He never addresses why it is that a silicon-based system cannot be conscious either, as all his arguments are circular on this front.

Less sophisticated critics bash the arguments that silicon-based systems could be intelligent or conscious by pointing out that an equivalent computer could be built out of beer cans, in which each bit of information could be represented by whether a particular dedicated beer can was currently upright or upside down. Indeed such computers could be built and would be able to do precisely the same computations as any silicon computer could do, albeit thousands of millions of times slower. The argument then dismisses a set of beer cans as being intelligent as patently obvious, and so demolishes silicon computers' being intelligent. Like Searle's Chinese Room, there is no real argument made against a beer can computer being intelligent—mere ridicule is used. Sort of like the idea that the world can't possibly be round because everyone in Australia would fall off. Ridicule does not make a valid argument. Ridicule instead of reason is a well-known refuge for tribalism.

Really then, most people's arguments about whether a robot can ever be conscious or have emotions are driven by people's own emotions, by their own tribalism. As Damasio argues, our reason is driven by our emotions. And that is true even when we are reasoning about our emotions and those of our machines.

Ultimately the arguments come down to core sets of unshakable beliefs. My own beliefs say that we are machines, and from that I conclude that there is no reason, in principle, that it is not possible to build a machine from silicon and steel that has both genuine emotions and consciousness.

Is There Something Else?

Although I have derided the use of "new stuff" in attempts to explain how we are different from machines, we are still left with a dilemma, and I will now use my own "new stuff" argument to hypothesize a way out of it. Even worse, I will not supply any data to back up this argument—exactly my criticism of both Penrose and Chalmers.

We know that our current robots are not as alive as real living creatures. Although Genghis can scramble over rough terrain, it does not have the long-term independence that we expect of living systems. Although Kismet can engage people in social interactions, eventually people get bored with it and some start to treat it with disdain, as though it is an object rather than a living creature.

Are there other fields of endeavor where we are able to really make artificial systems act like real biological systems? Beyond robotics, there is another field, artificial life, or Alife, in which biology is modeled, and, as in robotics, we have made tremendous progress in building systems with interesting properties. But as with robotics, there is well-founded criticism that so far the systems are not as robust, and ultimately as lifelike, as real biological systems.

In Alife people have built systems that reproduce, inside a computer-based world. Back at the beginning of the nineties, Tom Ray, then a biologist at the University of Delaware, developed a computer program called Tierra. This system simulated a simple computer so that Ray could have complete control over how it worked. Multiple computer programs competed for resources of the processing unit in the simulated computer. Ray placed a single program in a 60,000-word memory and let it run. (His words were only five bits, to approximate the information content in three base pairs of amino acids where there are only twenty proteins coded for plus a little control.) The program tried to copy itself elsewhere in memory and then spawn off a new process to simultaneously run that program. In this way the memory soon filled with simple "creatures" that tried to reproduce by copying themselves.

Like the DNA in biological systems, the code of the computer program was used in two ways. It was interpreted to produce the process of the creature itself, and it was copied to make a child program. The simulated computer, however, was subject to two sources of error. There were "cosmic rays" that would occasionally randomly flip a bit in memory, and there were copying errors where, as a word was being written to memory, a random bit within it might flip. Thus as the memory of the simulated computer filled with copies of the original seed program, mutations appeared. Some programs simply failed to run anymore and were removed. Others started to optimize and get slightly smaller, because the scheduling policy for multiple programs implicitly had a bias for smaller programs. Before long, "parasites," less than half the size of the original seed program, evolved. They were not able to copy themselves, but they could trick a larger program into copying them rather than itself. Soon other sorts of creatures joined them in the soup, including hyperparasites, and social programs that needed each other to reproduce.

When Tom Ray presented this work at an Alife conference in Santa Fe in 1991, there was great excitement. It seemed as though this were the key to building complex lifelike systems. Instead of having to be extremely clever, perhaps human engineers could set up a playground for artificial evolution, and interesting complex creatures would evolve. But somehow a glass ceiling was soon hit. Despite many years of further work by Ray and others, and experiments with thousands of computers connected over the Internet, nothing very much more interesting than the results of the first experiments have come along. One hypothesis about this was that the world in which the programs lived was just not complex enough for anything interesting to happen. In terms of information content, the genomes of his programs were four orders of magnitude smaller than the genome of the simplest self-sufficient cell. Also, the genotype—the coding—and the phenotype—the body—for Ray's creatures were one and the same thing. In all real biology the genotype is a strand of DNA, while the phenotype is the creature that is expressed by that set of genes.

A few years later Karl Sims came along and built an evolution system in which the genotypes and phenotypes where different. His genotypes were directed graphs with properties that allowed easy specification of symmetry and segmented body parts, such as multisegment limbs. Each element of the phenotype expressed by these genotypes was a rectangloid box that might have sensors in it, actuators to connect it to adjacent rectangloids, and a little neural network, also connected to the neural networks of adjacent body parts. A creature was formed out of many different sized rectangloids, coupled together to form legs, arms, or other body parts. The specified creature was placed in a simulated three-dimensional world with full Newtonian physics including gravity, friction, and a viscous fluid filling the volume. By changing the parameters of the fluid it could act like water, air, or a vacuum.

Sims had about a hundred creatures simulated in a generation. They would be evaluated on some metric, or fitness function, like how well they swam or crawled, and then the better ones would be allowed to reproduce. As with Ray's system, there were various ways in which mutations could happen. Over time the creatures would get better and better at whatever they were being measured for. The first sorts of creatures Sims evolved were those that could swim in water. Over time they would evolve either into long snakelike creatures or into stubbier creatures with flippers.

When these were placed on dry ground, they could not locomote very well, but by selecting for creatures that were better at locomoting, soon they evolved to be better and better at it. Some interesting results occurred along the way. Sims soon found he had to be very careful about exactly what fitness function he chose, or evolution would often come up with something he was not expecting. An early version of the locomotion fitness function did not penalize vertical motion and was only applied to a few seconds of existence. Soon really tall creatures evolved that were good at falling over and even tumbling. They scored high marks for the amount of motion they did in a few seconds of simulated time, but they were not really locomoting in any sustainable way. Then for a while creatures evolved that moved along by beating their bodies with their limbs—they had evolved to take advantage of a bug in the implementation of conservation of momentum in the simulated physics.

Eventually, Sims set the creatures a task of trying to grab a green block placed between two of them in a standardized competition format. Many different strategies quickly evolved. Some creatures blindly lashed out and tried to scoop up whatever was in the standard position with a long articulated arm. Others were more defensive, quickly deploying a shield against the opponent and then grabbing the block at their leisure.

Others locomoted toward the block and tried to run off into the distance, pushing the block in front of them, scooped in by a couple of stubby arms.

This work was very impressive and got people excited. It reignited the dreams inspired by Ray's work that we would be able to mindlessly evolve very intelligent creatures. Again, however, disappointment set in, as there has not been significant improvements on Sims's work in over five years.

Recently, Jordan Pollack and Hod Lipson implemented a new evolution system with similar capabilities to Sims's program. Besides evaluating their creatures in simulated Newtonian physics, they also took the extra step of connecting their creatures directly to rapid prototyping fabrication machines. Their creatures get manufactured physically in plastic, with links and ball joints. A human has to intercede to snap in electric motors in motor holders molded into the plastic. Then the creature is free of cyberspace and is able to locomote in the real world. This innovation has once again inspired people, but there is still no fundamental new idea that is letting the creatures evolve to do better and better things. The problem is that we do not really know why they do not get better and better, so it is hard to have a good idea to fix the problem.

In summary, both robots and artificial life simulations have come a long way. But they have not taken off by themselves in the ways we have come to expect of biological systems. Let us assume that all the Penroses, Chalmers, and Searles are wrong. Why is it then that our models are not doing better?

There are a few hypotheses we could make about what is lacking in all our robotic and Alife models.

1. We might just be getting a few parameters wrong in all our systems.
2. We might be building all our systems in too simple environments, and once we cross a certain complexity threshold, everything will work out as we expect.
3. We might simply be lacking enough computer power.
4. We might actually be missing something in our models of biology; there might indeed be some "new stuff" that we need.

The first three of these cases are all somewhat alike, although they refer to distinctly different problems. They are all alike in that, if one of them turns out to be true, it means that we do not need anyone to be particularly brilliant to get past our current problems—time and the natural process of science will see us through.

The first case, getting just a few parameters wrong, would mean that we have essentially modeled everything correctly but are just unlucky or ignorant in some minor way. If we could just stumble across the right set of parameters, or perhaps get a little deeper insight into some of the problems so that we could better choose the parameters, then things would start working better. For instance, it could be that our current neural network models will work quantitatively better if we have five layers of artificial neurons rather than today's standard three layers. Why this should be is not clear, but it is plausible that it might be so. Or it could be that artificial evolution works much better with populations of 100,000 or more, rather than the typical 1,000 or less. But perhaps these are vain hopes. One would expect that by now someone would have stumbled across a combination of parameters that worked qualitatively better than anything else around.

Now consider the second case. Although there was disappointment in the ultimate fate of the systems evolved by Ray and Sims it was the case that the environments the creatures existed in did not demand much of them. Perhaps they needed more environmental pressure in order to evolve in more interesting ways. Or perhaps we have all the ideas and components that are needed for living, breathing robots, but we just have not put them all together all at once. Perhaps we have only operated below some important complexity threshold. While this is an attractive idea, many people have been motivated by the same line of thinking and it has only so far resulted in systems that seem to suffer from "kitchen sink" syndrome. So again, while this is possible I rank it as having low probability of being the only problem.

The third case, that of a lack of computer power, is nothing new. Artificial intelligence and artificial life researchers have perennially complained that they do not have enough computer power for their experiments. However, the amount of computer power available has continued to follow Moore's law and doubled every eighteen months or two years since the beginning of each field. So it is getting hard to justify lack of computer power as the thing that is holding up progress. Of course, there has been major progress in both fields, and many areas of progress have been enabled by the gift of Moore's law. And sometimes we have indeed seen sudden qualitative changes in the way systems seem to function, just because more computer power is available.

We have recently seen an example of this. After being defeated by Deep Blue, Garry Kasparov said that he was surprised by its "ability to play as though it had a plan and how it understood the essence of the position." Deep Blue was no different in essence from the earlier versions he had been playing in the late 1980s, and in fact was not very different from Richard Greenblatt's program of the mid-sixties. Deep Blue still had no strategic-planning phase, as other chess programs designed to model human playing had. It still had only a tactical search. Because of the amount of computer power it had, it had a very deep, fast tactical search. Whereas Greenblatt's program was restricted to looking at a few hundred positions per second, Deep Blue was able to look at 200 million per second. The result was a program that seemed to Kasparov to have a game plan, not because there was anything new, but because more computer power made the approach feel qualitatively different. Bob Constable at Cornell University has reported to me the same sort of qualitative change of feeling he has recently noticed as he watches his automated theorem-proving programs at work. He *knows* that there is nothing qualitatively new, but he says that as he observes his programs operate, the deeper searches now made by his programs give behavior that feel to him like they have much more strategic plans on how to prove theorems.

These are two good examples because in both cases we know that the details of how the programs work, to play chess and to prove theorems, are nothing like the way humans attack these same problems. There is simply not enough of the slow hardware we have in our brains to look at as many positions or propositions as these programs do. But out of that mindless set of rules comes something that appears to two non AI researchers—world leaders in their respective intellectual fields—to be reasoning and thought.

Perhaps the same can happen for all that we hold dear about our humanity. Perhaps our current models of intelligence and life will become intelligent and will come to life if we can only get enough computer power. I am still doubtful that more computer power, by itself, will be sufficient, however.

I think we probably need a few Einsteins or Edisons to figure out a few things that we do not yet understand. I put my intellectual bets on the fourth case, above, that we are still missing something in biology, that there is some "new stuff."

Unlike Penrose, Chalmers, and even Searle, however, I am betting that the new stuff is something that is already staring us in the nose, and we just have not seen it yet. The essence of the argument that I will make here is that there is something that in some sense is obvious and plain to see but that we are not seeing it in any of the biological systems that we examine. Since I do not know what it is, I cannot talk about it directly. Instead, I must resort to a series of analogies.

First, let us use an analogy from physics, and building physical simulators. Suppose we are trying to model elastic objects falling and colliding. If we did not quite understand physics, we might leave out mass as a specifiable attribute of objects. This would be fine for the falling behavior, since everything falls in Earth's gravity with an acceleration independent of its mass. So if we implemented that first, we might become very encouraged by how well we were doing. But then when we came to implement collisions, no matter how much we tweaked parameters or no matter how hard we computed, the system would just not work correctly. Without mass in there, the simulation will never work.

So far, this analogy is a little like Chalmers's argument. My next step will depart from Chalmers however. Chalmers's argument calls for this missing thing, mass in my example above, to be something additional to the current physics of the world. He calls for it to be disruptive of our understanding of the universe. Analogies for what Chalmers hypothesizes in terms of how it would rock the current scientific world would be the discovery of X-rays a century ago, which led ultimately to quantum mechanics, or the discovery of the constancy of the speed of light, which lead to Einstein's theories of relativity. Both these discoveries added completely new layers to our understandings of the universe. We eventually realized that our old understanding of physics was only an approximation of what was really happening in the universe, useful at the scales at which it had been developed, but dangerously incorrect at other scales.

My version of the "new stuff" is not at all disruptive. It is just some new mathematics, which I have provocatively been referring to as "the juice" for the last few years.[2] But it is not an elixir of life that I hypothesize, and it is does not require any new physics to be present in living systems. My hypothesis is that we may simply not be seeing some fundamental mathematical description of what is going on in living systems. Consequently we are leaving out the necessary generative components to produce the processes that give rise to those descriptions as we build our artificial intelligence and artificial life models.

We have seen a number of new mathematical techniques arrive with great fanfare and promise over the last thirty years. They have included *catastrophe theory, chaos theory, dynamical systems, Markov random fields,* and *wavelets,* to name a few. Each time, researchers noticed ways in which they could be used to describe what was going on inside living systems. There often seemed to be confusion about the use of these mathematical techniques. Having identified these ways of describing what was happening inside biological systems using these techniques, researchers often jumped, without comment, to using them as explanative models of how the biological systems were actually operating. They then used the techniques as the foundations of computations that were meant to simulate what was going on inside the biological systems, but without any real evidence that they were good generative models. In any case, none of these ended up providing the breakthrough improvements that the early adopter evangelists had often predicted.

None of these mathematical techniques has the right sort of systems flavor that I am hypothesizing. The closest analogy that I can come up with is that of computation. Now I am not saying that computation is the missing notion, rather that it is an analogy for the sort of notion that I am hypothesizing.

First, we can note that computation was not disruptive intellectually, although the consequences of the mathematics that Turing and von Neumann developed did have disruptive technological consequences. A late-nineteenth-century mathematician would be able to understand the

2. When I first expounded this theory at a workshop in Switzerland, I was forty years old. At dinner that evening a young graduate student from Oxford told me that what I had said was very interesting and that he thought a lot of people came to similar sorts of ideas when they were in the sunsets of their careers.

idea of Turing computability and a von Neumann architecture with a few days instruction. They would then have the fundamentals of modern computation. Nothing would surprise them, or cause them to cry out in intellectual pain as quantum mechanics or relativity would if a physicist from the same era were exposed to them. Computation was a gentle, nondisruptive idea, but one that was immensely powerful. I am convinced that there is a similarly, but different, powerful idea that we still need to uncover, and that will give us great explanatory and engineering power for biological systems.

For most of the twentieth century, scientists have poked electrodes into living nervous systems and looked for correlations between signals measured and events elsewhere in the creature or its environment. Until the middle of the century this data was compared to notions of control of cybernetics, but in the second half it was compared to computation; how does the living system compute? That has been a driving metaphor in scientific research. I am reminded that, early on, the nervous system was thought to be a hydrodynamic system, and later a steam engine. When I was a child, I had a book which told me that the brain was like a telephone switching network. By the sixties, children's books were saying that the brain was like a digital computer, and then it became a massively parallel distributed computer. I have not seen one, but I would not be at all surprised to see a children's book today that said that the brain was like the World Wide Web with all its cross references and correlations. It seems unlikely that we have gotten the metaphor right yet. But we need to figure out the metaphor. In my view it is likely to be something like a mathematical formalism of something of which we can currently see all the pieces, but cannot yet understand how they fit together.

As another analogy to computation, imagine a society that had been isolated for the last hundred years and that had not invented computers. They did, however, have electricity and electrical instruments. Now suppose the scientists in this society were given a working computer. Would they be able to reduce it to a theoretical understanding of how it worked—how it kept a database, displayed and warped images on the screen, or played an audio CD, all with no notion of computation? I suspect that these isolated scientists would have to first invent the notion of computation, perhaps spurred on by correlations of signals that they saw by taking measurements on visible wires, and even inside the microprocessor chip. Once they had the right notion of computation, they would be able to make rapid progress in understanding the computer, and ultimately they would be able to build their own, even if they used some different fabrication technology. They would be able to do it because they would understand its principles of operation.

Now we have the analogies for the mathematics that I suspect we are missing. But where might we look for such mathematics?[3] Ah, if only I knew! It is certainly the case that living systems are made up of matter, and that our current computational models of living systems do not adequately capture certain "computational" properties of that matter. Real matter cannot be created and destroyed arbitrarily. This is a constraint that is entirely missing from Alife simulations of living systems. Furthermore, all matter is doing things all the time that would be incredibly expensive to compute. Molecules are subject to forces from other molecules around them, and physics operates by continually minimizing these forces. That is how the shapes of the membranes of cells come about, how molecules migrate through solution and across barriers, and how large complex molecules fold up on themselves to take the physical shapes that are important for the way they interact with each other as the mechanisms of recognition, binding, or transcription.

But this is just one obvious place to look. The real trick will be to find the nonobvious, for if the juice hypothesis is true, that must be where it is hiding.

It might turn out that, for all the different aspects of biology that we model, there is a different juice that we are missing. For perceptual systems, say, there might be some organizing principle, some mathematical notion, that we need in order to understand how animal perception systems really work. Once we have discovered this juice, we will be able to build computer vision systems that are good at all the things they currently are not good at. These include separating objects from the background, understanding facial expressions, discriminating the living from the nonliving, and general object recognition. None of our current vision systems can do much at all in any of these areas.

Perhaps other versions of the juice will be necessary to build good

3. Robert Rosen, a mathematical ecologist, posits that we may have to generalize our current understanding of physics to understand life, and makes a case for *category theory* being the crucial mathematical tool.

explanations of other aspects of biology. Perhaps different juices for evolution, cognition, consciousness or learning, will be discovered or invented and let those subfields of AI and Alife flower.

Or perhaps there will be just one mathematical notion, one juice, that will unify all these fields, revolutionize many aspects of research involving living systems, and enable rapid progress in AI and Alife.

Clever People and Aliens

Thus my version of the "new stuff" argument boils down to wanting us to come up with some clever analysis of living systems and then to use that analysis in designing better machines. It is possible that there is indeed something beyond such an analysis, that there really is some "new stuff" that relies on different physics than we currently understand. We have seen this happen twice in just over the last one hundred years, first with radiation, which ultimately lead to quantum mechanics, and then with relativity. Sometimes there really is new physics, and if we use that term broadly enough, it must surely cover what it is that makes things alive.

The next concern is whether people will ever be clever enough to understand the "new stuff," whether it is my weak version of just being better analysis, or whether it really does turn out to be some fundamentally new physics. What are the limits of human understanding?

Patrick Winston at MIT, in his undergraduate artificial intelligence lectures, likes to tell a story about a pet raccoon that he had when growing up in Illinois. The raccoon was very dexterous, able to manipulate mechanical fasteners and get to food that was supposed to be inaccessible. But Patrick says that it never occurred to him to wonder whether some day raccoonkind might eventually build a robot raccoon, just as capable as themselves. His parable is meant to be a caution to MIT undergraduates that perhaps they are not as smart as they like to think (and they certainly do like to think that way) and that perhaps we humans have too much hubris in our quest for artificial intelligence.

Could he be right? The downfall of Roger Penrose is that he refuses to accept limitations on the abilities of human mathematicians, and that leads him to misinterpret Gödel's theorem. Perhaps there is even some superversion of Gödel's theorem that says that any life-form in the universe cannot be smart enough to understand itself well enough to build a replica of itself through engineering a different technology. It is a bit of a stretch to think about the formal statement of such a theorem, but not difficult to see that in principle it might be the case that such a thing is true.

If such a thing is true, it still leaves the door open for someone smarter than us to figure out how we work, and to build a working machine that is just as emotional and just as clever as us. In what way could someone else be intrinsically smarter than us? This again is an issue of not wanting to be not special.

Let us consider someone who is not quite as smart as all of us but does not realize it. Todd Woodward et al., psychologists at the University of Victoria in British Columbia, report on a patient, called JT, who suffered a brain hemorrhage. As with many such people, JT was left with some mental functions intact but others impaired. In JT's case he suffered from a *color agnosia,* a disruption of his ability to understand colors. His disruption, however, is most interesting for how mild it is. By comparing JT to normal control subjects Woodward and his coworkers were able to ascertain that JT had normal performance in distinguishing colors from each other. In abstract color gratings he was able to count various bars distinguished only by changes in color and not brightness. So called color-blind people are very poor at this task. JT was also able to manipulate color words, successfully verbally filling in the color names in phrases like "green with envy." So he is able to see different colors and use color words. The problem for JT comes when he must associate these two different abilities. When asked to name color patches on a screen, JT got only about half of them right, most typically getting similar colors confused—e.g., yellow and orange, or purple and pink. When asked to use color pencils to color in line drawings of familiar objects, JT was able to get about three-quarters of them right. These experiments were carried out, and some were repeated, over a period of years. Thus, JT was not improving his abilities, he was permanently damaged.

At first blush, this seems very strange indeed. Someone who can "see" colors and can use color words but cannot make the right associations between them. Surely we would all be clever enough to figure things out and simply relearn the colors? But if that is so, why can JT not? He was a functioning adult with a technical job before his hemorrhage. He was still a functioning adult person afterward, but had a deficit that the rest of us can see. How many "deficits" do the rest of us have, but that all of us in our land of the blind just do not see?

With small wiring changes in their brains, people can have strange deficits in what they can reason and learn about. As the product of evolution, it is unlikely that we are completely optimized, especially in cognitive areas. Evolution builds a hodgepodge of capabilities that are adequate for the niche in which a creature survives. It is entirely possible that with a few additional wiring changes in our "normal" brains we would have newfound capabilities. These could be newfound capabilities that we cannot currently reason about, just as with the agnosia patient. They would be capabilities that our own special reasoning, of which we humans are so proud, is not capable of reasoning about, without the wiring already in place.

As we relate to chimpanzees and macaque monkeys and their intellectual abilities, we can imagine, at least in principle, a race that has evolved with pretty much all the capabilities of our brains, but with additional wiring in place, and perhaps even with newer modules. Just as some of our modules have capabilities that are not present in chimpanzees, a *supersapiens* might have modules and capabilities that are not even latently present in us. Rather than being from Earth, perhaps supersapiens is from another planet, orbiting one of the billions of stars in one of the billions of galaxies that populate the universe. Now what will happen when supersapiens look at us? Will they see a raccoon with dexterous little hands? Will they see an impaired agnosia sufferer ("agnostic" just does not seem like the right word here) who simply cannot reason about things that are obvious? Or will they see a race of individuals capable of building artificial creatures with capabilities similar to their own?

The Question of Consciousness

Suppose that we are able to build machines in the not-too-distant future that we all agree have the emotions of a dog, say. Suppose further that the robots we build like this can act like a dog, and are as good companions for us as dogs. What then will we say about whether they are conscious or not? The question of consciousness is of great concern to Penrose, Chalmers, and Searle, and indeed many think that this is the key to understanding what makes us human.

This is a difficult question, for in general we cannot agree on whether any animals but us are conscious at all. While there is a gradation of emotional content, and hence empathy, which we attribute to different animals, there is no such consensus among people about the consciousness of animals.

Part of the problem is that we have no real operational definition of consciousness, apart from our own personal experience of what it is like to be us. We know that we ourselves have consciousness, and by extension we are willing to grant it to other human beings. But there is always a gnawing worry that perhaps their experience of consciousness is not the same as our own experience. Perhaps we, our own selves, really are unique, and all those around us, despite their verbal reports, are not experiencing consciousness in the same way we do.

Once we extend this questioning to other animals, to orangutans, to dogs, to mice, to birds, to lizards, and to insects, we get progressively less sure of just how much consciousness they possess. Some people like to deny any hint of consciousness to any animals but ourselves. It becomes much more acceptable to slaughter whales for meat, or scientific research, or whatever happens to be the current rationale, if they are totally unconscious. It is also comforting to our sense of specialness if we alone are conscious.

In my opinion we are completely prescientific at this point about what consciousness is. We do not exactly know what it would be about a robot that would convince us that it had consciousness, even simulated consciousness. Perhaps we will be surprised one day when one of our robots earnestly informs us that it is conscious, and just like I take your word for your being conscious, we will have to accept its word for it. There will be no other option.

Ethical Slaves?

One of the great attractions of robots is that they can be our slaves. Mindlessly they can work for us, doing our bidding. At least this is one of the versions we see in science fiction.

But what if the robots we build have feelings? What if we start empathizing with them. Will it any longer be ethical to have them as slaves? This is exactly the conundrum that faced American slave owners. As they or their northern neighbors started to give humanhood to their slaves, it became immoral to enslave them. Once the specialness of European lineage over African lineage was erased, or at least blurred, it became unethical to treat blacks as slaves. They, but not cows or pigs, had the same right to freedom as did white people. Later a similar awakening happened concerning the status of women.

Fortunately we are not doomed to create a race of slaves that is unethical to have as slaves. Our refrigerators work twenty-four hours a day, seven days a week, and we do not feel the slightest moral concern for them. We will make many robots that are equally unemotional, unconscious, and unempathetic. We will use them as slaves just as we use our dishwashers, vacuum cleaners, and automobiles today. But those that we make more intelligent, that we give emotions to, and that we empathize with, will be a problem. We had better be careful just what we build, because we might end up liking them, and then we will be morally responsible for their well-being. Sort of like children.

Flesh and Machines

Further Reading

Brooks, R. A. 2001. "The Relationship Between Matter and Life." *Nature* 409: 409–11.

Chalmers, D. 1996. *The Conscious Mind*. New York: Oxford University Press.

Lipson, H., and J. B. Pollack. 2000. "Automatic Design and Manufacture of Robotics Lifeforms," *Nature* 406: 974–78.

Maturana, H. R., and F. J. Varela. 1987. *The Tree of Knowledge: The Biological Roots of Human Understanding*. New Science Library. Boston: Shambhala Publications.

Nolfi, S., and D. Floreano. 2000. *Evolutionary Robotics*. Cambridge, Mass.: MIT Press.

Penrose, R. 1989. *The Emperor's New Mind*. New York: Oxford University Press.

———. 1994. *Shadows of the Mind*. New York: Oxford University Press.

Ray, T. S. 1991. "An Approach to the Synthesis of Life." In *Artificial Life II*. Edited by C. G. Langton. Redwood City, Calif.: Addison-Wesley.

Rosen, R. 1991. *Life Itself*. New York: Columbia University Press.

Searle, J. R. 1992. *The Rediscovery of the Mind*. Cambridge, Mass.: MIT Press.

Sims, K. 1994. "Evolving 3D Morphology and Behavior by Competition." *Artificial Life* 1: 353–72.

Woodward, T. S., et al. 1999. "Analysis of Errors in Color Agnosia: A Single-Case Study." *Neurocase* 5: 95–108.

9.   Them and Us

Within twenty years our computers will be a thousand times as powerful as they are now. Many authors have compared the amount of computation that could be going on in a human brain to how much is going on inside a standard personal computer. While it could be that everyone's measure is completely wrong in some way we just do not understand, at the moment there does seem to be a consensus that we have it right. That said, the amount of computational power in a personal computer will surpass that in a human brain sometime in the next twenty years.

The chance that we might have robots that are more intelligent than we are then becomes a real possibility, with a few caveats. If it turns out that the "juice" of chapter 8 is something noncomputational, then this may not come to pass. And even if it is computational, it still may not come to pass until we have the requisite number of Einsteins working in this field. And if it turns out that humans are like the raccoons of chapter 8, we still may not ever be able to build machines more intelligent than ourselves. Unless we turn out to be just smart enough to set up an artificial evolution system that produces the right programs. If, by the way, the juice is noncomputational, it may or may not be that we are able to build the appropriate technology to implement it. So we may or may not get to produce a robot more intelligent than ourselves, or it might happen in as few as twenty years.

What if we do? Many people have thought about the consequences of such robots, and in general there have been two sorts of prominent views of what the future may hold: one is damnation, and one is salvation.

Damnation

Damnation has long been the prediction of Hollywood and some fiction writers. One could argue that the theme was first revealed in Mary Shelley's *Frankenstein,* where life is given to something previously not alive, and it engenders great fear in the masses. But other early writing on robots tended to emphasize their familial or erotic connotations. It was really modern science fiction that turned robots into killers.

The great Arthur C. Clarke in collaboration with director Stanley Kubrick succumbed to damnation in *2001: A Space Odyssey*. HAL 9000, conflicted by its orders, eventually decides that the mission is more likely to succeed if the human crew on board the space ship is eliminated. Ultimately, HAL is foiled by its lack of actuation capability as the last astronaut, Dave Bowman, is able to knock out its higher intelligence centers. But in the meantime it has killed three hibernating crew, and one space-walking astronaut—sent out on a bogus mission by HAL.

Isaac Asimov, the other great *hard-science* science fiction writer of the twentieth century deliberately protected humankind from his robots by insisting that they were built with the three laws of robotics that we discussed in chapter 4. While he never exactly explained how it was that society managed to enforce across all of humanity that all robots be built in this way, the effect of the laws was to make robots nonthreatening to humans, at least on the surface. Many of Asimov's stories explored that which ensued from a situation set up such that the laws were in logical conflict. Often the robots faced moral dilemmas over how they should act, and which transgression of which law was the greater moral evil. Other, deeper stories explored the limitations on the humanity of robots that were imposed by the adoption of these laws. The robots in general had a feeling of being second-class and quested for ways in which they could become more human. The recent movie adaptation of an Asimov story, *The Bicentennial Man,* mentioned earlier, explored exactly this question—ultimately the robot protagonist gives up life in order to die as a human.

But in movies both pre- and postdating 2001 such as *Colossus: The Forbin Project, The Matrix,* and *The Terminator,* the story is less subtle and rather different. The robots or supercomputers decide to take over the world from humans because they are seen by the computers as intellectually inferior and unreliable.

The premises of some of these movies are more believable than others, but all require a fairly severe suspension of disbelief in order to enjoy watching the movie as entertainment. A problem arises when otherwise sensible people get confused by the premises of the movies and the future of our society. Surprisingly, clever technologists and humanists who managed to make it through viewing *The Exorcist* and *Poltergeist* without taking up a public cry against the dangers of witchcraft have recently had difficulty watching Hollywood "robots take over the world" movies without publicly fretting about the future of the human race.

Under the Hollywood "robots take over the world" scenario, humans create new life-forms that develop over time. They grow smarter and smarter, and before too long think of themselves as smarter than their creators. They start to squabble with their creators and ignore their advice. They ultimately part ways from their creators and establish themselves as separate entities.

Up to this point there is no difference between the Hollywood script and the normal process of people raising children. In the latter case the young adults and their parents eventually come to some mutual understanding and both continue through life together more or less, and happy more or less. In the Hollywood version the robots or computers created by consequence-blinded technologists have no human feelings and ultimately decide the world will be simpler by getting rid of their human precursors.

There are many hidden assumptions that are necessary for this to all work out. I will list a few of them here and then discuss the time frame in which they might come about.

- The machines can repair and reproduce themselves without human help.
- It is possible to build machines that are intelligent but which do not have human emotions and, in particular, have no empathy for humans.
- The machines we build will have a desire to survive and to control their environment to ensure their survival.
- We, ultimately, will not be able to control our machines when they make a decision.

Reproduction. As mentioned in chapter 8, a couple of researchers have recently had artificial evolution systems automatically fabricate, using rapid prototyping machines, examples of the evolved locomotion machines. The publication of this work caused a worldwide media stir, as many reporters jumped to the conclusion that self-reproducing machines were just around the corner. The machines that were built, however, were rather simple locomotion machines, with only structural members being built automatically. The motors and wire, which are much more complicated, were not built automatically. Furthermore, the machine that did the building of the structural members was much more complicated than the constructed robot. A self-reproducing robot or robot society might be able to rely on some infrastructure external to themselves, just as we humans rely on plants growing and synthesizing complex proteins for us to eat and thus fuel our activity and reproduction. But we do have to tend that infrastructure, so ultimately any self-reproducing robots will as a society also have to maintain and produce the machines, such as the rapid prototyping machine, that they need. This is a complex requirement and beyond the current state of the art. However, these considerations pale when it is remembered that the crawling robots that were automatically manufactured relied on a conventional computer to control them. For serious reproduction, robots would need to reproduce their brains, their computers. The central processing chip for a modern computer is built in a fabrication facility that costs well over $1 billion. Although they are some of the most highly automated facilities on Earth (to avoid contamination from hair and skin particles falling off people, if nothing else) there are still many thousands of people involved in keeping such a plant supplied and running. Our robots will need to master all of these tasks, or somehow we and they will need to come up with a smaller infrastructure to do the same task. Just as paper did not disappear from our offices as promised in the early nineties, one can safely assume that chip production will involve a lot of human intervention for the next twenty-five years.

No emotions or empathy. Our children who ultimately take over from us as societal leaders and inherit our wealth are rarely cold-blooded about getting rid of us in the way the fictional robots take over from and eliminate humans. Our children have love and empathy for us. Their emotional bond to us, and indeed to all of humanity, is what makes our self-reproducing society work. As we get more sophisticated in our understanding of the world, we as a society have increased our range of empathy, extending it at an almost human level both to the great apes and to whales and dolphins. At the same time, we have come to understand how critical emotions are for all intelligent animals in the world—how emotions control and shape the rational decision-making process. We have built emotional machines that are situated in the world, but not a single unemotional robot that is able to operate with the same level of purpose or understanding in the world. To build emotionless robots that can operate in the world intelligently at the moment requires us to encode every fact about the world directly for them. There is no indication that such an approach to artificial intelligence is near to fruition. I do not see it making much progress in the next twenty years; in the meantime the emotion-based intelligent systems for robots situated in the real world are plunging ahead. These robots will not hate us for what we are, and in fact will have empathetic reactions to us.

Survival instinct. To date, we have not built robots with any strong survival instinct, but this is for lack of trying rather than any technological capacity. Some people have suggested building robots that maintain their own energy supplies. There is a project under way at the University of the West of England to build a robot that harvests slugs from fields, takes them to a stationary digestion system, and retrieves energy made from burning methane produced by decomposing the slugs. Such robots are not likely to become practical in the near term—they are a long way off from escaping an energy deficit at this point. A more short-term version of an energy self-sufficient robot would be one that deliberately seeks out wall sockets and plugs itself in to recharge whenever it finds one. This is technically possible today, although no one that I know of has built one.[1] Such robots would rely on the operation of the electrical power supply, and could soon be rendered powerless, literally, by switching off the electricity supplies. We are many decades off from robots that can metabolize any other source of energy in a way that will let them live off the land and get away from our control of their energy sources.

Loss of control. As long as we control the energy sources for our machines, we will certainly control the machines. We would have to deliberately put in a system without fail-safes for us to lose control of even energy–self-sufficient robots. Or we would have to build machines that could deliberately outsmart us. The latter are not likely to appear overnight—just as evolution took some time to hone the systems it built, with incremental improvements, balancing all the constraints and just taking small steps at a time. We will know when we are close to building machines over which we will have no control, as before that we will have machines over which we have little control, and before that

1. There was a machine, called the Hopkins Beast, built at the Applied Physics Lab at Johns Hopkins University in the mid-sixties that was able to plug itself into a wall socket, but it was not able to hunt down sockets with any generality—it was able to plug itself in to only its one test wall socket.

we will have machines over which we occasionally lose control. Such machines will not appear overnight as the work of one brilliant lone scientist. If they were to, we should surely be worried, as that loner might be mentally deranged and might just for the fun of it build an uncontrollable robot. But just as it took many people working for many years to get from the Wright brothers' *Flyer* to a modern Boeing 747, it will take many tens of thousands of people working together to get from where we are today with our robots to one that we could not control. The Russians in the movie *Dr. Strangelove* deployed a "doomsday bomb" that was set to automatically explode and destroy all human life should anyone anywhere set off a nuclear explosion. Those fictional characters knew exactly what they were doing, and so will we, should we choose to build an uncontrollable robot. It is unlikely to be possible within the next fifty years, and it is unlikely to ever be a desirable social goal.

There is no reason *in principle* why we should not be able to build machines that satisfy our four premises above. The question will be whether we, or someone else, wants to.

Should we, perhaps, adopt Asimov's three laws for all the robots that we build? People have often asked me whether the robots in my laboratory are built according to Asimov's laws. My answer, as in chapter 4, is always no, not because I do not wish my robots to obey those laws, but rather because up until now I have not known how to build them in that way. The problem has always been the level of perceptual abilities required by robots to be able to obey the laws.

All the laws require that a robot be able to perceive human beings and to differentiate them from other things in the world. This has been fairly difficult up until very recently but is now becoming possible, and even commonplace. But beyond that, even the first law gets very difficult to implement. In order to avoid harming people the robot would have to have an understanding of the consequences of its actions. That is still difficult for any robot, because it needs a full perceptual model of the world in order to do that. Understanding when its inactions could lead to a person getting hurt require it to have a model of human behavior and be able to predict what will happen to humans as they go about their business. By the time we get to the third law, things get very complicated. Should it drop an object when its gripper motor is about to burn out? By the third law, yes. But what if the object is sharp and a person is nearby and has their hand underneath the object right now? The robot has to be able to keep track of people's body parts. We expect a human adult to be able to obey these laws, at least at the surface level, when they are not in conflict. But in order to do so the person has to have a very rich and deep understanding of the world—beyond that of any of our robots today.

We may be able to put something like Asimov's laws into our robots in the future, at least in those robots that are smart enough to understand them. If they are not that smart, then we do not have to worry about them taking over from us. Like a missile treaty or nuclear nonproliferation treaty, the nations of the world might decide that all robots should have some restrictions on them. The impetus for this may well come from wanting to reduce the horrors of war—just as humanity has collectively decided to draw back from the specter of biological weapons. One might expect within the next ten years that we will start to see a push toward robot treaties, so that robots, for instance, might be banned from being offensive weapons. Just where cruise missiles will qualify under such a ban will be an interesting question. While Asimov's laws themselves are both too complex and too vague to be used directly (despite the delicious irony that this would present of a self-fulfilling prophecy of Asimov's), some similar sort of statutes might well be implementable and enforceable.

Salvation

An alternative view is held by many researchers in robotics, artificial intelligence, and computer science. They think that intelligent robots will provide a path to immortality. They expect that before they die the technology will be available that will let them download their consciousness into a computer or robot and they will be able to live in this form forever.

Hans Moravec has been the earliest and most vocal proponent of this possibility. That is the same Hans Moravec with whom I shared an office at Stanford when he was working on the Cart, described in chapter 2. But there have been plenty of other people, such as Marvin Minsky and

Ray Kurzweil, who have succumbed to the temptation of immortality in exchange for their intellectual souls.

Hans and Ray both point to Moore's law as the critical factor in changing our relationship to machines. They rightly point out that no matter how much computation one estimates is going on inside a human, that will soon be surpassed by the computation available in a desktop machine. They assume that we will therefore soon have human-level equivalence in our computers and robots. They are perhaps being slightly optimistic here—while both of them have worked on artificial intelligence technologies for the last thirty years, neither is able to give a prescription for what new insights, apart from just the sheer amount of computer power, that will get us to human intelligence equivalence.

The simplest version of the Moravec-Kurzweil salvation is that these intelligent computers and robots will provide us with unimaginable wealth brought on by the fantastic levels of productivity that the new technology will provide. If history is any guide, this is unquestionably true. We in the Western world live in levels of comfort that were unimaginable to royalty just a few centuries ago. We have abilities, and finances, to travel in ways that nobility could not imagine even a century ago. Our standard of living continues to rise by all objective measures. There is now some debate, however, whether our subjective standard of living is continuing to rise—are we more stressed or less stressed than we were ten years ago? Be that as it may, the introduction of intelligent robots into our everyday lives will surely continue the objective increases in our standard of living. But some people take this good news a little too far and foresee a complete utopia where robots are at our beck and call as we all write poetry and eat grapes by the bunch. The truth will probably be neither so fantastic nor so boring.

The more complex salvation scenario involves people leaving their mortal bodies. This is fairly common among salvation scenarios in most religions, so it is not surprising to find it in techno-utopia. The idea is that once we have wonderful robotic bodies, and plenty of silicon computation on board, we will be able to make a simulation of our brains that runs in the computer to control the robot. Other versions have us entirely disposing with a body, even silicon and steel, and simply living in the ethereal realm of cyberspace. This is the more sophisticated version of salvation.

But how do we squeeze ourselves into silicon? We do it by taking apart our brain, piece by piece, and simulating it in computation. Moravec's version of this has a team of surgical robots slicing away little pieces of our brain at a time, and building a simulator for each neuron as it is dissected. The simulated neuron gets coupled back into the living brain, and our consciousness is uninterrupted. Then the next neuron is physically dissected and it too is added to the simulation. After a mere million million such steps (Hans never mentions how long it is going to take) a virtual version of our brain is running in the computer.

This strong version of salvation seems plausible in principle. However, we may yet be hundreds of years off in figuring out just how to do it. It takes computational chauvinism to new heights. It neglects the primary role played by the bath of neurotransmitters and hormones in which our neuronal cells swim. It neglects the role of our body in placing constraints and providing noncomputational aspects to our existence. And it may be completely missing the *juice*.

Ray Kurzweil would argue with me about my pessimism on the time scale to solve all these problems. He is predicting a singularity in which computation makes us all-powerful around the year 2020. After this eschatological cataclysm (as Jaron Lanier calls it) we will be able to push through technological roadblocks at previously unimaginable speeds. Of course, the year 2020 is right around when Ray will turn seventy years old himself. And he is determined to live at least that long—he has written a book on low-fat diets and how to avoid heart disease. If he can make it to 2020, he can make it to anywhere.

In 1993, I attended a technology and art conference, "Ars Electronica," in Linz, Austria, where my former postdoctoral student Pattie Maes gave a talk titled "Why Immortality Is a Dead Idea." She took as many people as she could find who had publicly predicted downloading of consciousness into silicon, and plotted the dates of their predictions, along with when they themselves would turn seventy years old. Not too surprisingly, the years matched up for each of them. Three score and ten years from their individual births, technology would be ripe for them to download their consciousnesses into a computer. Just in the nick of time! They were each, in their own minds, going to be remarkably lucky, to be in just the right place at the right time.

A variation on this theme, and a backup, is to have oneself frozen in liquid nitrogen should you die. The idea is that in the future, *sometime,* medicine will have progressed far enough that it will be possible to cure whatever it was that caused one's death and to reverse the process of dying, as long as the person was frozen very soon after death. For more than thirty years there has been a California-based industry freezing technological free thinkers who have paid enough cash up front. A number of people mentioned in this chapter are signed up to be frozen when they die. Even if their physical brain itself cannot be revived, the thought goes that when it is commonplace to be able to download people's consciousnesses into computers, at least that much will be retrievable from frozen human flesh.

Originally the cryogenics companies froze whole bodies, letting them sit in a container of liquid nitrogen. Later, as expenses mounted, there was a little technological revisionism. If medicine in the future is going to be good enough to revive a whole person, brain included, surely it will be good enough to be able to fabricate a mere body somehow. So a whole bunch of people were removed from their private nitrogen swimming pools, and their bodies were amputated. Just the heads went back into much smaller, cheaper containers. If actual revival rather than downloading is what ultimately happens, there are going to be a number of shocked heads waking up bodiless. Those who are not yet frozen seem to have accepted this in stride, however, and are quite happy to just have their heads frozen.

When the bodies were amputated, there was some concern. Many of the cells had completely burst during the freezing process. The task of restructuring a thinking human mind is going to be awfully difficult, much more so than originally thought. But since we are relying on future technology, this is no big deal—the frozen heads will just have to stay frozen for a little longer, until technology is ready to deal with them.

Now there is one little problem with all this. Will future generations be interested in reviving frozen heads from the late twentieth and early twenty-first centuries? In a very crowded world, just how many under-educated, socially backward, technologically incompetent people will it be worth reviving? A few here and there, especially the famous ones who have special memories, might be worth reviving for historians to chat with. But they will not quite fit in the future world. Imagine if we were able to revive the 5,000-year-old frozen man found in the Italian Alps in the early nineties. It would be incredibly interesting to do so, in order to find out more about the times in which he lived. And for that benefit we would probably be happy to take care of him for the rest of his natural revived life—even if he never was able to quite fit into modern society. But suppose we found 1,000 other such frozen individuals, or 10,000, or 1 million. And suppose each one was incredibly hard to revive with years of dedicated effort from a large team of scientists. Would we really want all of them around? Would we bother with them all?

I am a little worried about all my friends who are going to have themselves frozen. They may not turn out to be very important or worthwhile to future generations. They may not return even if technology makes it possible.

Downloading to computers and frozen heads are part of a more general belief system, popular among many technologists, known as transhumanism, or extropianism. The goal of extropians is to extend human lifetimes through technological innovations (some of what they hope for will be discussed in the next chapter). Much of what they hope for will come to pass, but a lot of what they hope for is driven by a fear of death. The history of human religions warns us about the passions, mysticism, and irrational beliefs that our human brains can accommodate when driven by a fear of death. Perhaps the only thing worse than not being special is not being alive. Humans will go to great lengths to avoid an acceptance of death, of our personal mechanism grinding to a halt. So I am careful to try to separate the beliefs of extropians into those that are driven by technological imperatives and those that are driven by the fear of the unknown.

We are a long, long way from being able to download ourselves into computers or robots. While in principle it will ultimately be possible, it is not a worthwhile place to look for personal salvation for those of us who are alive today. We are going to have to find some other way to come to peace with our own personal futures. I think we are all going to have to die eventually. Just like all those humans before us. Technology is not going to save us yet.

Where To?

It may be that we are not smart enough to build self-reproducing intelligent robots. Or it may be that the juice really is something beyond mere physicality and that it is fundamentally impossible that such robots might ever exist.

While I worry about the first possibility, I think the second possibility is as unlikely as its turning out that the Earth really is flat after all. The last 2,500 years of our science and technology have led in the direction that the universe is understandable in purely physical terms. We have progressively been able to give up the need to use spiritual mechanisms to explain the motion of objects, the motion of the planets and stars, chemical processes, and biological processes. Thinking otherwise is most likely just clinging to our desire to be special. We need to have clear heads and face up in an intellectually honest way to something erroneous that evolution has bred into us to enable us to be tribal and brutal and to survive on the savanna of nonspecialness.

The question then is when, not if, we will build self-reproducing intelligent robots. And when we have, be it twenty years from now, or a thousand, will it be damnation or salvation? Or could there be a third path?

The Third Path

There are alternatives to the scenarios of intelligent self-reproducing robots. The null alternative is that not much is going to happen that is different from the past. The foreseeable future will be much like the recent past. The very increasing pace of innovation and change that we have all experienced will just continue, perhaps slowing down to some manageable constant rate of change, and nothing much structurally will change.

Under this alternative the third millennium will be rather like the second, but with even better plumbing.

But this belies the change that has happened in the world over the last fifty years. A billion people now use airplanes with some frequency. We travel distances many times a year that are farther than those that the most adventurous used to travel in a lifetime. The more elite of us travel to places halfway around the world (as far as you can go) a number of times a year, in less than twenty-four hours of travel. When I was teaching on Wednesdays and Fridays in Massachusetts, it was not uncommon for me to squeeze in a trip to Japan over the weekend and not miss class at all.

After some grainy initial satellite broadcasts in the mid-sixties, we were stunned by seeing man land on the Moon, live in our own living rooms. Since then we have more and more come to expect to see everything of importance anywhere in the world or in space, as it happens at the press of a hand-held remote button. We expect to see inside every courtroom and in every bedroom. We expect to have fifty or two hundred video channels delivered to our homes around the clock. And we expect now to be able to tap into thousands more, somewhat selectively, on the Internet. And we are now about to embark on transmitting ourselves via remote presence to be able to participate in remote action.

At the start of the twentieth century, if we did travel, we expected nothing more than an occasional letter from those we had left behind. Now we think nothing of telephoning or e-mailing around the the world many times per day. We expect instant contact in our personal lives—we carry cell phones and can reach our loved ones instantly whether they be out shopping or on a business trip on the other side of the globe.

We have also become accustomed to immediate access to the world's information online, twenty-four hours a day, without regard to its source or its physical location. I search the white pages in Australia, browse research papers on a Japanese computer, and check the schedule at the Louvre, all from my wireless laptop, sitting on my living room couch. I stumble upon things that I myself wrote fifteen years ago in databases in Germany. I search the scientific literature from the last fifty years, looking for relevant papers for things that I need to know accurately to write this book. Information is global.

The world reacts in a global way to that global information. Stock markets are correlated as the sun sweeps across the globe. There is a globalization of our culture because trends and ideas can travel across the globe almost instantaneously. When we colonize Mars, there will be only a few tens of minutes' lag across all of humanity even when we are on opposite sides of the Sun. Perhaps it is time to seriously start colonizing our recently discovered neighboring solar systems to restore the evolutionary imperative of isolated populations.

Technology has changed our relationship with time and space. The end of the second millennium was vastly different from the end of the first in this regard, and even unimaginably different from just fifty years before. Computers and communication, following quickly on the footsteps of the jet airplane, have changed the structure of our world. It has changed so quickly that even our science fiction cannot keep up. Episodes of *Star Trek* made early in the year 2001, and supposedly set four hundred years hence, had people physically carrying information from one part of a star ship to another in handheld data pads. We no longer do that in our homes or offices. Data travels on networks between rooms and floors (perhaps with a minor side trip a thousand miles away to some Internet hub) in the blink of an eye.

The last fifty years have seen a profound change in our relationship to technology. Everything we do in our lives, from turning the key in the ignition, to making coffee, to talking on the phone, is mediated by computers and computation. Hence the fear struck in our hearts by the Y2K bugs that were lurking in so many places, but which were mostly eliminated in time. Our whole global economy, down to almost every economic transaction in which we participate, is reliant on computation. The physical tokens that had been developed over more than 2,000 years of commerce have largely been eliminated in just fifty years. There is no going back now. We have made a pact with technology that lets us live in such vast numbers with such a high standard of living.

These transitions have changed the way we view our place in the world and in the universe. We are the masters of the world rather than being an element of the landscape. We are surprised when nature bests us, when a storm closes down our cities, or a forest fire rages out of our control. The normal order of things is that the world should operate as we wish it, and on our time scale. It never quite does, but somehow our expectation is that it should. Sometimes it may feel to us that nothing really fundamental has changed in the last fifty years, but in fact it has. Our technology has let us change our attitude about our place in the world. This is as significant a change as was the Renaissance, or the industrial revolution. We and our world are different.

After such a tumultuous change in just fifty years, we might think that it is time for a breather. But the majority of people who have ever worked in science or technology are alive today. There is no chance that technology will not continue to change our existence in a significant way.

While we have come to *rely* on our machines over the last fifty years, we are about to *become* our machines during the first part of this millennium. We need not fear our machines because we, the man-machines, will always be a step ahead of them, the machine-machines. We will not download ourselves into machines; rather, those of us alive today, over the course of our lifetimes, will morph ourselves into machines.

Further Reading

Kurzweil, R. 1990. *The Age of Intelligent Machines*. Cambridge, Mass.: MIT Press.

———. 1999. *The Age of Spiritual Machines*. New York: Viking.

Moravec, H. P. 1988. *Mind Children*. Cambridge, Mass.: Harvard University Press.

———. 1998. *Robot*. New York: Oxford University Press.

Regis, E. 1990. *Great Mambo Chicken and the Transhuman Condition*. Reading, Mass.: Addison-Wesley.

10.   Us as Them

As director of the MIT Artificial Intelligence Laboratory, I have the pleasure of telling visitors about all the far-out fantastic work that is being done in our lab. Not only am I lucky enough to be at perhaps the world's premier technology institution, but I am able to represent a laboratory at that institution which is at the extreme of creativity and, in the information technology area, the hottest technology of our time. Now of course at MIT there is also a downside. There happen to be two other even bigger information technology laboratories at the same place, competing for attention with the same sponsors. In 1997, *U.S. News & World Report* ranked the MIT Laboratory for Computer Science, the MIT Media Laboratory, and the MIT AI Lab as three of the country's top-ten information technology laboratories. So the cloud in the silver lining is that I often have to share the innovation limelight with our sister laboratories. While feeling to us like just about one of the best labs in the world, there are two siblings on the same block to keep us humble about our own accomplishments.

The Media Lab had a big impact during the nineties with their wearable computers. A number of students, including Steve Mann and Thad Starner, started wearing computers permanently. They usually had a single-hand keyboard in one hand, and a video screen covering one eye, so that as they walked around campus, sat in classes or meetings, or went about their research, they were always connected and computing. They were integrating computers into their everyday life in a completely pervasive way. And they were doing it visibly. Before long they became known as the Cyborgs. Part human and part machine, they had an identity of their own on the MIT campus. Colin Angle and I had experimented with some of these ideas in late 1989, and we had even dressed Colin up in some mocked-up interfaces for a series of concept photos. I had taken the idea to a conference in Japan, where I was the last-minute replacement for Arthur C. Clarke when he was too ill to travel, and there I talked about all sorts of future work that might be done. But we had never done it, and the Media Lab had instead done their own take on the idea. The AI Lab did not have any cyborgs. I often wondered, what if I had pursued those ideas, what if . . .

One day late in 1999, I walked from my office on the ninth floor of the AI Lab building out to the elevator lobby. The freight elevator from the basement stopped and the doors slid open. Out walked a modified version of Hugh Herr. Hugh has a half-time appointment as an assistant professor at the Harvard Medical School and a half-time appointment as a research scientist in the MIT AI Lab. As Hugh stepped from the elevator, a chill went up my spine. From the thighs up he was all human. From the thighs down he was all robot. And not an elegant robot. He was prototype robot. Metal rods instead of bones, computer boards where muscles would normally be, batteries hanging on by black electrical tape, and wires dangling everywhere. Now *this* was a cyborg!

Hugh is a double-leg amputee, and one can see how that has shaped his professional career. He completed a Ph.D. with the late Tom

McMann at Harvard on animal locomotion. He worked with Gill Pratt in the AI Lab on legged robots, and then the two of them teamed up to develop a robotic prosthetic leg that is now in production for amputees everywhere to use. More recently he has started work on using cultured mammal muscle to actuate small robots and aims to eventually build artificial legs in this way—biological muscles rather than electric motors.

Hugh's motivations are plain for all to see, and already his work has helped many other people who are also amputees. His tenacity is admirable. The tenacity of others working to replace other sorts of lost human functionality is equally strong. With solid clinical credentials and motivations, they are developing a new set of technologies that will soon be adopted into human bodies wholesale, in ways the inventors did not originally imagine.

We the Machines

Today there are tens of thousands of people walking around with implants that connect electronics directly to their nervous systems. These people have accepted that they are better off becoming a hybrid, part human, part machine, than staying purely human. I am speaking here of people with cochlea implants that enable them to hear.

As with most of the technologies I am about to discuss, there are good, understandable clinical reasons why those with damage to the small hairs in their cochleas have chosen an artificial augmentation of their body. They have made the choice because it restores to them an ability that they have lost.

For people who grew up able to hear and who learned to hear and speak a language in the normal way, but later lost hearing because of damage to their cochlea, an implant is often able to successfully restore their ability to hear spoken language. The implant is an electronic device that separates multiple sound frequencies received by a microphone in the ear and outputs the strength of six or so frequencies on electrodes. The electrodes are implanted next to nerve cells that would normally receive signals from sensor cells that transduce the motion of the small cochlea hairs. They are lined up so that the frequencies measured by the artificial cochlea are inserted along the length of the real cochlea at the place that would normally be responding to approximately the same frequency.

With these implants people are able to hear well enough to understand speech better than less impaired people using a hearing aid only. The small number of frequency bands does not at this point let people hear music or other sounds very well, but that ability may come as more sophisticated implants are built over time—there are very active research programs in both the United States and Europe in building better implants.

The devices are surgically inserted into people's ears, and electrodes are permanently implanted, so that there is direct electrical connection between the electronics of the silicon device and the nervous system of the patient. They hear through a combination of flesh and machine.

Artificial cochleas are not the first machine components to be inserted in human bodies on a very large scale. For years people have used structural implants, ranging from plates and screws to support broken bones, to wholesale replacements of hip joints. Indeed, a hip-joint–replacement operation is hardly remarked upon today. Most people in the Western world have relatives or friends with artificial hips. There are also chips that are regularly implanted in animals in many parts of the world. In the United Kingdom all dogs have a chip implanted under their skin. This can be interrogated externally so that their identity can be established against a national registry—a dog that shows up at the pound can be returned to its owner whether they want it or not. Understandably there have been a handful of advocates and a bevy of opponents arguing about whether the same sort of thing should be done for all people at birth. In any case, these chips are not really part of the body of the dogs in which they are installed. They operate completely independently from the dog, without any interface with its neural system.

But artificial cochleas are not, in fact, the first implants that introduce electrical activity into people's bodies. There have been heart pacemakers implanted under people's chests for over thirty years. A small periodic electrical signal stimulates heart muscle, and the heart entrains that stimulation, enabling it to beat regularly. Such an electrical connection somehow seems more mindless than that of the artificial cochleas. The cochleas process sensory information, partition it, and send it to multiple destinations in the nervous system. The artificial cochlea really becomes an integral part of the way in which a person with one implanted is able to sense the world.

Many people are working to improve the performance of the artificial cochleas. They are increasing the number of frequency channels handled, they are increasing the sophistication of the signal processing done on the sounds that are received, and they are looking at better ways to implant the electrodes into the patient's flesh. All these improvements are responses to the natural demands of the medical marketplace. There are many people who lose their hearing due to natural degeneration of their ears, through damage from sustained loud noise, or through infections. Providing such technology and administering it to people is just like any other form of clinical medicine. There really seems to be very little in the way of a moral argument that could say that using technology in this way is bad. As such, we can expect it to continue.

While improvements are being made in cochlea implants, work is progressing in many countries toward the first retinal implants. These are for previously sighted patients who have damaged retinas. The most common target is for people with macular degeneration. This is a progressive disease that attacks the foveal part of the retina and gradually takes away people's ability to see fine detail. Those affected lose the ability to read and recognize faces, and eventually they can do no more than get general contextual cues from their peripheral vision.

A retinal implant is a silicon retina, like the pixel array inside a video or digital camera. The idea is to collect light in the electronic pixels, then send that information to the spatially appropriate nerve cell, or neuron, in the optic nerve that would have received such information from the retina itself had it still been intact.

Retinal implants are much more complex than cochlea implants. Rather than just a handful of electrodes, it will be necessary to interface with the nervous system at tens of thousands of different nerves or neurons. This increases the complexity of both the device and the implantation technique.

The way in which sound is processed by an intact cochlea is well understood and reasonably straightforward to emulate in electronics.

The processing the retina does is not so well understood at this time. Thus, although the spatial structure of how light falling onto a particular part of the retina should be mapped to which neurons is easy to understand, there is still a lot of research necessary to understand just what signals should be transmitted. These unknowns, along with the increased number of connections necessary, mean that artificial retinas are not as far along as artificial cochleas. There are no people walking around today with artificial retinas permanently implanted. They have been experimentally implanted in volunteers for day-long periods. The subjects report being able to distinguish light and dark and notice "differences" when looking at different things. They can hardly be said to *see* in the way that cochlea implant patients can hear, however. Nevertheless, progress is being made. It seems reasonable to expect that artificial retinas will become clinically expedient in the next decade or so. People who can be helped by them are a natural market, and helping blind people to see is a noble goal.

There are many other arenas where there is a clinical push to connect silicon circuits directly to nervous systems.

Many people have suggested making neural connections between nerves in the stump of an amputee to a prosthetic arm or leg. For legs this turns out to be not such an attractive option. People need their legs to adapt as they walk over rough ground, smooth ground, up stairs and down stairs, but their legs do not need to be really dexterous. The possible payoff from having direct neural connections to control their artificial legs is not great enough to compensate for the myriad inconveniences and problems that it brings. They want to simply strap their leg on in the morning and not need to be worried about plugging in connections, whether they be through a connector extruding from their flesh, with its associated infection problems, or a carefully placed receiver for skin galvanic currents, or a wireless connection from an implant within the leg stump. Gill Pratt and Hugh Herr have accommodated this desire by inventing an artificial leg that can tell what the person is doing and adjust its actions accordingly. While a doctoral student, Ari Wilkenfeld programmed the first prototype prosthetic leg to interpret sensors and decide whether the person was walking up stairs or down stairs, or across relatively flat terrain, and also to sense how fast the person was trying to walk. The intelligent leg mitigates the need for direct neural connection.

Artificial arms are another story, however. People want to be able to do many different things with their arms. They want to be able to grasp many different-sized and -shaped objects with different requirements of delicacy and force to be used. They want to be able to pick up and put down objects, but they also want to be able to turn the page of a book, or put the stopper in the kitchen sink, or turn a door handle, or pull the refrigerator door open, or open an envelope, or turn the ignition key in their car. They want to be able to blow their nose and wipe their bum.

For prosthetic arms there is a real need for high bandwidth connections between the wearer's nervous system and the artificial device. To this end there are many experiments with animals in developing ways to have permanent connections between electrical devices and neurons. The most promising way seems to be to implant a silicon chip with holes through it right in the path of a severed cluster of nerve cells. The nerves regrow through the holes, and the silicon circuits can both measure electrical activities in the nerves and inject their own signals into the nerves. The chip communicates wirelessly to an external monitor strapped to the skin. There are many research issues that still need a lot of work. What sort of material should be used so that it neither harms the body nor is rejected by the immune system? How should the signals on the nerves be interpreted? How can the brain of the person best be trained to interpret the signals, such as those that might come from force or touch sensors in the artificial arm?

This work is proceeding. There are clinical pushes for it to be made to work, and there are thousands of people who will be very grateful when it becomes practical. But there is not a total urgency for those with one missing arm. They are able to function in the world, though not as well as they would like, and deserve. For people who have damage to their spinal cords, especially high up on their spinal cords, the situation is much more immediate and dire. Their whole life is defined by their inability to move. Some people with really high-up damage are not able to breathe on their own even, and the only part of their bodies they can control is their eyes. They cannot talk, just look. They have immediate and pressing needs for new technologies that let them connect their brains to robots.

There are some things that can be done without direct neural connections. One of my graduate students, Holly Yanco, modified a wheelchair and put in a robot control system. The wheelchair is able to follow corridors and go through doorways on its own indoors. Outdoors, it is able to follow along a sidewalk, and avoid falling down stairs. Holly used a system developed by Jim Gips at Boston College that allows the profoundly impaired to give signals to a computer by moving their eyes. Small electrodes taped to the temples and forehead pick up the electrical signals that their brain sends to the eye muscles and the computer can infer which ways their eyes have moved. Holly hooked this up to her robot control system. A person riding in the wheelchair simply looks at an icon on a screen attached to the wheelchair's arms, such as GO FORWARD, and the robot wheelchair takes care of the second-to-second adjustments in speed and direction necessary for successful navigation.

While such aids can help the very disabled, they themselves clearly want something better. There have been some experiments with a handful of very disabled patients in making direct connections between their nervous systems and a computer. One patient is able to move a mouse pointer about on his computer screen just by thinking. There is an implant into his nervous system and wires coming out of his body that connect him to a computer interface card. With this small attachment the patient is able to have infinitely more control over his own life than he could before. He can type and send messages to people. He can read their messages as he chooses, and he can surf the Internet, bringing the information resources of the world into his service, displaying what he wants before his own eyes.

Experiments such as this are very limited. There are ethical questions about what sort of invasive technologies should be put inside a human patient when so little is known about how well it will work. So experiments proceed with animals, but often it is hard to know just what capabilities an implant gives them. As reported in November 2000, however, there was a recent experiment with monkeys that had a very clear outcome.

Miguel Nicolelis at Duke University implanted electrodes into the motor control region of the brains of small new-world owl monkeys. Over a period of two years they monitored the signals on these electrodes as the monkeys moved their arms about to reach for food. By putting the food in different locations relative to the monkeys, the researchers were able to get the monkey to do the same motion many times. The electrodes picked up signals from many hundreds of neurons. With some signal-processing techniques they were able to measure the electrical activity in these individual neurons and watch that activity during different arm reaches to different places. After collecting and analyzing enough data the researchers were able to predict the activity in individual neurons for any particular reaching task that they gave to the monkey.

Ordinarily such predictions followed by analysis of recorded data would have been enough to establish that the researchers had found a correlation between neural activity and particular motions of the monkey's arms. But Nicolelis and his colleagues came up with a much more vivid demonstration. They attached a real-time computer to the electrodes and programmed it to predict where the monkey's arm was about to move. Then they connected that computer to a robot arm and had it move in the direction and reach that it predicted the monkey was about to undertake. The monkey reaches for a piece of food, and the robot arm, unseen by the monkey, moves in the same direction and reaches out the same distance. As an encore, the research group at Duke teamed up with a group at MIT and had a robot arm in Massachusetts reaching simultaneously with the monkey in North Carolina. Monkey remote presence. Almost. The monkey never knew about the existence of the robot arm.

The follow-on set of experiments from this work are to give the monkey feedback from the robot arm, so that it can feel the forces and touches that the robot arm encounters in the world. This sort of force and haptic feedback has become routine in the robotics world, but the signals are usually reflected as forces on a person's arm. With the direct brain connection of the Duke monkeys the leap will be to send the forces directly into the monkeys' brains. This work will not be simple. Figuring out how to represent the forces and touches electrically, and which nerve cells they should go to, will be a challenge. Since perceptual signals go through multiple processing stages in the brain, there will be multiple candidate sites. Which ones will give the monkeys a sense that the robot arm is really theirs will be a difficult question. The place that the signals are injected will have to be one where the adaptive mechanisms of the brain are able to operate, to map and learn the correlations between sensation and what is really happening in the world. We know from the "phantom limb" phenomenon—where someone who loses an arm both adapts to the loss and yet still can feel the nonexistent arm—that this is a complex issue. Thus we do not yet know how difficult it will be to really have the monkeys feel that the robot arm is their own.

Such work is critical to being able to attach prosthetic arms to human amputees and to let their brains control the arms directly, while feeling that the arm is an extension of themselves. Likewise, this work is necessary to give quadriplegics the ability to control robots as substitutes for their own bodies. At the moment there appear to be no insurmountable obstacles to making this work out. There is research to be done, and the details of what will be necessary are far from clear. But the current indications are encouraging that it will not be too many years before such work will be ready to be tried on human patients.

The monkeys in these experiments live carefully circumscribed lives. They are treated well, but they are not subjected to the complications of a life in the wild, a life in a rough-and-tumble monkey society. These experimental animals have electrodes in their brains, sticking out through their skulls, so that computers can be hooked up to them directly. This will not be the way amputees and quadriplegics will wish to interface to their technologies. In the short term there will need to be extremely low-powered wireless connections implanted directly in their brains so that they can communicate through their skulls and skins with the devices that are external to their bodies. Later, when artificial arms are permanently attached to their bones and their own skin merges with the skin of the prostheses, it will be possible to route the wires from their brains down into the arms, and even the legs.

Routing of wires within people's nervous systems is another area that is an active area of research. Electrical stimulation of nerve cells in the thalamus has been used as an effective way to relieve the symptoms of Parkinson's disease. Sometimes the symptoms of this and other motor control diseases are so severe, making people shake so violently and consistently, that patients prefer to be surgically paralyzed in their affected limbs rather than live with the constant tiring motion. To make this happen it is necessary to cut major nerve fiber highways so that no signals can get through to muscles. Often the correct signals are present, but they are swamped by the other, spurious signals, caused by the disease, that are also present.

Some researchers are now looking into the possibility of bypassing diseased or damaged parts of the brain in cases of Parkinson's and other diseases, by routing nerve signals around those areas. By inserting wires into the brain and connecting them to nerve cells at either end, it is hoped that signals can be routed from motor control centers to muscles while the interfering natural pathways can be severed. This work is also in its early stages, but there are many suffering patients and good clinical reasons to pursue these possibilities.

Many of these technologies are going to come to fruition in the next ten years, and almost certainly they will all be perfected within twenty years.

The Future of Surgery

The nature of surgery is undergoing a rapid transformation. While human surgeons are still in charge, sometimes for good reason, and sometimes just for historical reasons, they are being augmented with computer vision and robotic aids.

Computer vision techniques that Eric Grimson and his students perfected at the MIT Artificial Intelligence Laboratory are used daily by surgeons at Brigham and Women's Hospital in Boston to remove brain tumors. The system measures the exact location of a patient and gives the surgeon X-ray vision inside the patient's head, overlaying MRI data that has been segmented and color-coded for the tumor and functionally different parts of the brain. The data is projected in such a way that as the surgeon looks at the patient through a special screen, or sometimes on a TV monitor on the side, he or she sees all this extra data exactly aligned with the patient. The surgeon can see things that were formerly not visible. The tools the surgeon uses are instrumented, and their exact three-dimensional position shows up in the displays too. Surgeries are now less invasive through smaller openings, and they take less time because the surgeons know more exactly what they are doing. These new "Nintendo" surgeons no longer watch their hands. They watch the TV screen as their nimble and skilled hands swiftly manipulate their tools.

The same techniques have been adapted for orthopedic and other surgeries. They allow for much less invasive surgeries than before because the surgeon does not have to directly see inside. He or she is given the virtual reality illusion of seeing inside, and that is enough to do the job well, even better than the limited view from just one set of eyes using the visual spectrum.

Robots themselves are also starting to be used for surgery, although again a human is in ultimate control. Intuitive Surgical of Mountain View, California, sells a remote-presence surgical system. The surgeon sits at a console with his fingers inserted into loops. As he moves his hands, a tiny robot with very different shape and form moves inside a human patient on the other side of the room. A camera inside the patient gives the surgeon a view of what is happening. As the surgeon twists his wrists and reorients his hands, the two small robot manipulators react accordingly. The surgical manipulators, developed at the MIT AI Lab by Ken Salisbury and Akhil Madhani, are able to cut and grasp. They are at the end of long rigid tubes less than half an inch in diameter that enter the patient through very small incisions. The tubes do not move at all during surgery, only the tiny manipulator at the end moves around inside the patient. The surgeon is able to elect to operate directly with 1 millimeter of their motion translating to 1 millimeter of robot motion, or at a five-to-one ratio, in which case the robot only moves a fifth of a millimeter. Intuitive Surgical's robots are being used throughout Europe and the United States for a variety of surgeries, including heart-valve replacements. So far, the surgeries have all been performed with the surgeon in the same location as the patient, but there is no reason that this could not change. Time lags across networks are ultimately determined by the speed of light, so it will not be possible to let the surgeons have effective control if they are too far from the patient, but distances of 100 kilometers or so should be possible in the not-too-distant future.

These sorts of surgical aids are just the tip of the iceberg. Many other robots are being developed for surgery throughout the world. For ethical reasons the work of all these robots is overseen by a human surgeon. But that may change in the future. For some of the systems the robot could carry out the motions themselves today. Humans are only in the loop to appease the fears of the patients and the overseeing hospital and regulatory boards. Just as we eventually gave up the demand that all automobiles be preceded by a human on foot waving a warning flag, we might eventually let these robots do more of the actual surgery. Twenty years from now it may be common to have surgeries overseen by medical technicians whose training will be years less than that of a surgeon. Surgery will become much more routine and common. Just as we now have laser eye surgery done in our shopping malls, correcting our vision permanently by changing the shape of our natural lenses, we will soon have other forms of elective surgery done in ways that are easily accessible to the masses.

Elective Surgery

Within twenty years surgery will be vastly more convenient and available than today, and there will be many techniques for embedding silicon and steel inside human beings to compensate for lost capabilities.

Of course, things will not stop there. There will be a whole new class of enhancements for our human bodies. These will all start innocently enough, but there will be inexorable pressure to push the technologies into more elective realms. Just as cosmetic surgery has become commonplace, technological body enhancements will become socially acceptable. People who are nominally healthy will start introducing robotic technologies into their bodies.

How could this happen? Will we really overcome our fear, and even revulsion, at changing our bodies into machines? Let us look at one scenario, for a gradual change that will lead to a wholesale reversal of attitudes.

Just as cochlea implants have become fairly routine, so too will retinal implants for people who have lost their sight through degeneration of their retinas. Initially people who have one good eye but one eye damaged through an accident will not be considered for implants. There will be sufficient demand from people with two bad eyes that it will be considered morally responsible to treat them first. But the demand from those patients will make the technology and operations more common and routine, and before too long people with one good eye will start to have the procedure on their bad eye.

But some people might opt for a little enhancement on their bad eye. Perhaps they would like a silicon retina that is enhanced for night vision. We already know how to build silicon arrays for digital cameras that are many times more sensitive than the human eye at night. So someone who has one good eye and has lived for many years blind in the other eye might decide that they can get around just fine during the day, but it would be great to be able to see at night—something that no ordinary person can do. Why not upgrade their useless eye to give them that capability? The silicon retina would need to have an electronic auto-iris so that it did not flood the person's optic nerve with signals during the day. In fact, in the early models it might have to simply shut down in daylight. But that is not something that might worry our hypothetical patient. Their bad eye is useless during both the day and night without the surgery.

Being able to see everything clearly at night is going to be an awfully interesting option for some people. Anyone who wants to operate clandestinely: soldiers, drug smugglers, terrorists.

Night vision enhancement will get to the point that some people with two perfectly good eyes may be willing to sacrifice one for it. In poorer countries people are already willing to sell some of their own organs for what appear to be pitiful amounts of money. In other parts of the world people are willing to become human bombs to support their causes. Modifying a good eye, to give superhuman performance, will not be too outrageous for lots of individuals, resistance movements, and governments.

People with extreme hobbies might also find eye modification useful and, when it becomes affordable, opt to do it. Mountain climbers, spelunkers, ultramarathoners, and Arctic trekkers and sledders might all be attracted by the idea of being able to see through the dark night. It would even help for driving a car at night. Our licensing authorities insist that we wear corrective lenses for driving if our eyes are not up to normal standards. Where does this slippery slope head, let alone end?

Night vision is not the only option, of course. A shift in the portion of the spectrum in which we can see might also be useful. More toward the infrared and we could become much more sensitive to seeing heat sources. This would be great for search-and-rescue teams, or even for firefighters. More toward the ultraviolet, and we could make all sorts of fine distinctions about the health of living plants—what farmer would not want to understand his crops better? And what if we could consciously change the sensitivities of our eyes, from night vision, to ultraviolet sensitivity, to infrared sensitive, back to normal? All we would need would be some connection between our neurons and the circuitry implanted in our previously normal eyes, and then the somewhat pesky detail of making sure that those neurons are under conscious control of the person in some intuitive way.

The clinical push on letting amputees and quadriplegics control external devices is not only the key to eventually being able to control our imagined modulatable eyes, but is also going to make us part of the Internet.

The current systems that let people control a computer mouse will let people control their eyes' spectral sensitivity. These techniques work by having the person imagine moving part of their body that they no longer have direct control over. Eventually that conscious effort becomes much more unconscious as the person's brain remaps their internal body image. We are all familiar with this phenomenon. When we first start driving an automobile, it is a thing that we are inside of and trying to control. Sometimes it feels like it has a will of its own and we need to concentrate to give the right control inputs to it. Before long, however, it becomes part of us. Initially as we drive into a parking spot, we are directing this box of steel to go to a particular place. After a while we are driving into the spot, and have a body awareness of the extent of our car as an extension of our body. It is no longer us and the car but just an extended us, with a car body.

There are other ways in which we extend ourselves too. Many of the technologies that have become indispensable parts of our lives are external to our bodies—they have become the new talismans that we carry with us everywhere we go. The most noticeable of these are our cellular telephones. We have become dependent on these to communicate with our families and our office. But now we are becoming dependent on them for all sorts of information services, ranging from weather predictions, train schedules, movie schedules, directions to places we are going, stock market prices, and purchasing objects both large and small.

Many of us also carry a personal digital assistant and have all of our professional lives scheduled on these devices, along with all our business contacts and our notes, drawings, and plans. Then, of course, there is the Internet, that external information space that dominates many of our lives. We gather, and post, most of the information that we use, and send and receive tens to hundreds of e-mail messages per day. We are chained to our desk machine, or our portable is glued to our lap (except during takeoff and landing) almost permanently so that we have access to these information channels. What if we could make all these external devices internal, what if they were all just part of our minds, just as our ability to see and hear is just a part of our mind?

A person with a thought-controllable mouse can browse the Internet by thinking, but that browsing is mediated by their eyes. Now combine the mouse with an implanted retina chip. Instead of having the artificial retina be a camera, make it a display device, connected to the computer that the thought mouse controls. Now the person could wander the information ways of cyberspace within a mental cocoon. But this would require the sacrifice of an eye. What if instead of inserting the visual image of the screen at the retina it was done in the rear of the brain in one of the visual-processing areas that reside there? When the display was switched off, everything in the visual system would work as usual. With the display switched on, there would be an interruption to normal service, with the screen image replacing what would normally have been seen. Now, there are quite a few details to work through to make all this work out as I have suggested. It may well take a solid twenty years of research and experiments to get there. But there does not seem to be an in-principle reason why this could not be made to work. It may take quite some practice and training for a person to be able to adequately perceive the appropriate information, but it certainly seems quite possible, and even probable.

Of course, it may turn out that there are better ways to interface the Internet and the equivalent of our PDAs to the insides of our heads. Rather than mediate all this information through a visual representation, somewhere in one of our retinotopic maps, it may eventually be possible to have the information appear much more directly in our minds. Most of us are pretty good at retrieving our home phone number, but beyond a dozen or two frequently dialed numbers, we need to go to an external device. When we think of our own number, we do not conjure up a visual image of the digits. Instead, the "number," whatever that means, is just there. When we go to our external device, we do see the number as visual images of the digits. When our external devices are surgically implanted in our brains, perhaps we will find a way to bypass that visualization step, and get the information directly.

Working out how to do this will be a significant research undertaking. However, there will be market pull. Initially there will be research pressure for the blind, with useless eyes, to have direct Internet access. That will drive the development of direct mental access to the Internet. Then, because the sighted majority will not need to sacrifice a good eye to get it, there will be plenty of pressure to allow them to get hold of this technology.

Once we have figured out this direct mental-tapping technology there may be a whole new set of services that spring up. Just as standard HTML Web pages proliferated, and then specialized WAP (Wireless Application Protocol) services for mobile phones with tiny screens came along, there might well be, twenty years from now, a whole host of "mentalese" service providers. They will be packaging information in a form most easily browsable with direct neural connections rather than optimizing it for visual presentation.

Of course, once there are Internet connections, all the services of cell phones and PDAs will be easily layered on top of this infrastructure. We will be able to communicate by thought with anyone else with the same technology implanted, anywhere in the world. Whether this form of communication will feel more like text-based instant messaging, or more like some sort of Vulcan mind-meld will depend on the particular technologies that it is possible to develop.

Having such things implanted in our brains will make us tremendously more powerful. Just as the current external Web and cell phones enable us to do more, more often, so too will our mental access to cyberspace. We will be able to think the lights off downstairs instead of having to stumble down in the dark to switch them off, and as an externally silent alarm goes off inside our head, leaving our spouse to sleep longer undisturbed, we will be able to think the coffee machine on, down in the kitchen.

Even face-to-face in a meeting, we might choose to open up a separate mental communication channel (perhaps ultimately transported by the cell-phone network, after all) with a particular person in the room, so that we can have a private side discussion with them, strategizing about how to proceed publicly. These, and many other as yet unimaginable capabilities, will change the very ways in which human beings interact. We will be superhumans in many respects. And through our thought-mediated connections to cyberspace, we will have access to physical control of our universe, just with our thoughts. The remote-presence robots of chapter 6 will be at our mental beck and call. While we may be physically present in one particular place, we will be able to mentally project ourselves to any remote-presence device and location for which we have authorization. Each of the have-nots will soon want to become one of the haves.

Initially the haves will be the weird ones. In our Olympic sports we do not let athletes who use performance-enhancing drugs compete. We might outlaw implanted Internet devices among students taking the SATs, as they will have an unfair advantage. But before long, just as with calculators, we might come to expect that everyone taking the SATs will have mental Internet access. What starts out as bizarre will probably become the norm.

Acceptability

Some people will be repelled by the idea of incorporating technology into their bodies, while others will be curious and even eager to try these modifications. Initially the choice will certainly be much more complicated than a personal decision. There will be strong issues of social acceptability. These issues will probably vary from country to country and even in different regions within a single country.

The acceptability of food made from genetically engineered crops is very different in different Western countries at the start of the new millennium. In Germany it is a very intensely debated issue, while it hardly registers in the United States. Likewise, the current issues around body modification are subject to very different social mores in different countries.

Kidney, liver, lung, and heart transplants are a current-technology form of body augmentation. For people with diseased internal organs transplants from a donor person are a replacement methodology. We currently do not have steel and silicon versions of these organs that work as well as biological organs. And unlike an eye transplant, say, such organs are able to make all the necessary connections when implanted in someone else's body and to function normally.

But accepting such organs into one's body is not a morality-free issue. For kidneys there is often the possibility of taking one from a willing living relative, since we all start out with two but are able to survive well with only one. In this case, we are accepting the flesh of another living person into our own body. For lung and heart transplants we are relying on the death of someone else, and then we have part of that dead person in our body, giving us life. Since livers can grow, even in an adult, both possibilities are open—a living or a dead donor.

Kidney and heart transplants are common in the United States. Indeed kidney transplants are now so common as to be unremarkable. In Japan, however, there have been only a handful of such operations in total. It was only in February 1999 that the second heart transplant was carried out. The first such transplant thirty-one years before had ended in accusations of murder against the transplant surgeon Wada Juro for removing a beating heart from the donor. In the West the notion of brain death has been long accepted, but it was only enacted as law in 1997 in Japan. The 1999 transplant from someone who had signed a donor card became a media circus because of the moral implications of removing organs from someone who was not "dead" by the traditional Japanese measures. Lest one think that this implies a consistently different Japanese sensibility for all human life, one must remember that, unlike in the United States, abortion is completely acceptable in Japan. And deformed thalidomide babies had three times the infant mortality rate that they had in the United States, despite much better overall infant mortality rates in Japan.

The point here is not that one community or country has better or superior moral values. Rather, different countries and communities hold to different moral values on particular issues. These moral values impact the acceptability of medical procedures even when life may hang in the balance for the recipient. They will certainly impact the social acceptability of cosmetic medical procedures such as silicon (as distinct from silicone) implants.

Such social constraints are not immutable over time, and we do see them change in relatively short periods on the order of a decade or two. The Japanese are easing their restrictions on transplants. Likewise, they are changing notions of the sanctity of the body and just what bodily modifications are socially acceptable. Pierced ears for women were taboo in Japan well into the nineties, but now a sojourn on the Tokyo subway shows that the taboo seems to have completely disappeared. In the West pierced ears on males were reserved only for pirates well into the eighties. Now a man with a pierced ear is unremarkable through most of the professional ranks in the United States. It would still be unlikely for a president or a Supreme Court justice to have such a body modification, but that too will change over time. Just as the baby boomers with their history of drug experimentation have gotten there, eventually those offices will be occupied by today's kids from generation Y, and there is no real going back on a hole through your flesh or cartilage.

Today many of us might say, "I don't want any stinking microchips inside my head, no titanium extensions to my bones, no sensory augmentations." It would feel unnatural. It would not be me. Some of us may change. Our children may feel differently. And their children almost certainly will.

Beyond Cyborgs

Our technology has been under development for thousands of years. It is just now getting to the point where we can incorporate it inside our bodies. And we will. We will change ourselves from being purely the product of our genetic heritage to a more Lamarckian sort of species wherein we will be the product also of our own technology.

For the moment that technology is based on silicon and steel. Before this new century is out, that technology will have long been surpassed, and our bodies by the middle of the century will reflect that next wave.

The robots of the mid twenty-first century will have silicon components, and steel components, and titanium, and maybe even some gallium arsenide, and certainly a bunch of other materials and super-conductors, and polymers, and structures that we have hardly even imagined. Our bodies too will contain all these technologies. But we and our robots are going to be full of a new round of technologies too—engineered biotechnology.

For fifty years we have been developing technology that lets us understand biology at the molecular level. Recently people have started turning that technology from analysis to synthesis. This is the standard transition from science to engineering.

The first attempts at using molecular biology technology as an engineering substrate have been brutish and clumsy but nevertheless extraordinarily powerful. Now work is proceeding on newer, more refined techniques that will be even more powerful.

Much work has been done in culturing cells and controlling their growth so that now a replacement pinna, the cartilaginous portion of the external ear, can be grown in vitro and attached permanently to a person who has lost one of hers. Experimental work also proceeds on growing replacement organs in vitro.

Recently there has begun serious work on using these sorts of technologies in robotics. Hugh Herr, the cyborg in the opening of this chapter, is not satisfied with the electrically operated dampers in his new generation of artificial legs. He wants active muscles but knows that electric motors will not have the right characteristics, and they will take large batteries. Hugh and his students at our lab have started building robots that are actuated by mouse muscles—muscles that can be grown in vitro from a single cell. His first robots must be bathed in a weak sugar solution to work, but they are the marriage of silicon and steel with biological matter. A small microprocessor receives high-level commands such as "swim," "turn right," etc., and turns them into coordinated signals that travel out on wires that innervate the biological but artificial muscles. The robot swims and turns right.

There are many research issues remaining, of course: how to shape the muscles appropriately as they grow, how to feed them sugar without their needing to be floating in it, and how to keep them alive over long periods so that a useful replacement leg can last a reasonable amount of time.

Hugh's work uses cells as they currently exist. But there is also work that modifies what goes on inside cells. There has been a lot of work done on genetic modification of existing organisms. The technology for inserting and deleting genes was critical for understanding the role of genes and for the roles of the proteins for which those genes coded. These techniques are now being used to insert new characteristics into crops or to remove genes that cause disease.

But such approaches are in some sense rather crude. They rely on mixing and matching existing genes to get some hoped-for result out of the complex dynamics of the interactions of all the proteins that are coded for by those genes. It is a little like plug and play with peripheral devices in your stereo system or your home computer. You are not really inventing anything new, just capitalizing on the ways in which the manufacturers have enabled all the components to communicate with each other. A less crude thing to do would be to build new components yourself. A really masterful thing would be to completely redesign the central processing unit of your computer. In the case of engineering genes, some work is starting on the middle ground—the complete redesign is still beyond us.

Tom Knight and Ron Weiss at the MIT Artificial Intelligence Laboratory have started to turn genetically engineered living *E. coli* cells into little tiny robots. They use a menu of sensors and actuators that exist already in the *E. coli*, or which they can easily genetically engineer in using the crude techniques that are well known. They have built little tiny *E. coli* robots that sense molecules (homoserine lactones, to be precise) that can be absorbed through the cell wall. They are also considering using naturally occurring sensors for pH, light, electric, and magnetic fields, and for other simple molecules. The actuators for their cell robots have been the same sorts of lactone molecules diffused through the cell walls, which emitted light, using a gene stolen from *Monocentris japonicus,* the Japanese pinecone fish. They are also looking into controlling flagellar motors, cell death, and the production of enzymes as actuators for these robots.

The most challenging part of Knight and Weiss's robots are the computations that they do in order to decide how to produce their outputs as a function of their current state and their inputs. They impose a digital discipline on these cells. They take simple digital circuits and compile them into a string of DNA that they insert in the genome of their population of *E. coli*. The molecular dynamics of the cells and the transcription mechanism of the cell is hijacked into doing the originally specified computations by this string of DNA. For instance, the compiler might choose to have protein A inhibit the transcription of protein B. By ensuring that there is the appropriate nonlinear transfer curve in this inhibition, the concentration of protein B in the cell can be seen as the logical inversion of the concentration of protein A. Thus the RNA transcription mechanism implements a logical NOT gate inside the cell, as determined by Knight and Weiss's compiler. More complex logic gates are easily built on top of this mechanism, and so complex computations can be forced to happen inside the living cell. With the right sensors and actuators a robot is born. Already Knight and Weiss have produced beakers full of robots, billions of robots of two different species, communicating with each other and switching on and off luminescence displays in response to messages from other robots and the concentration of signaling molecules in the solution. The computation speed of these robots is only impressive in that it is so slow—tens of minutes to make simple decisions. Such techniques will not replace silicon in our quest for ever more computation. What is important here is that computation ultimately controls some of the internal processes of a living cell.

In the not-too-distant future, people might achieve similar control over the molecular processes of living cells in more subtle ways, without going all the way to a digital process. But Knight and Weiss's work shows that it can be done. When we look ahead thirty years, we can imagine that we will be able to have programmed cells within living organisms, and even within ourselves. We will certainly have them in our robots, as they are very easy to manufacture. Living cells can already self-repair and self-reproduce. All you need to do is feed them simple sugars and you get more of them.

It is too early and the present too murky to see where all this leads, but it is clear that robotic technology will merge with biotechnology in the first half of this century. And so the robotic technology we are adopting into our bodies will ultimately become biotechnology: technology that will be programmed into our cells through modifications to our genes.

We are on a path to changing our genome in profound ways. Not simple improvements toward ideal humans as is often feared. In reality, we will have the power to manipulate our own bodies in the way we currently manipulate the design of machines. We will have the keys to our own existence. There is no need to worry about mere robots taking over from us. We will be taking over from ourselves with manipulatable body plans and capabilities easily able to match that of any robot.

The distinction between us and robots is going to disappear.

Further Reading

Knight, T. F., and R. Weiss. 2000. "Engineered Communications for Microbial Robotics." From Proceedings of the 6th International Workshop on DNA-Based Computers, DNA 2000, Leiden, The Netherlands. Edited by A.Condon and G. Rozenberg. In *Lecture Notes in Computer Science*. Vol. 2054, pp. 1–16. Berlin: Springer-Verlag.

Loeb, G. 2001. "Prosthetics, Neural." In *Handbook of Brain Theory and Neural Networks*. 2nd ed. Pp. 768–72. Edited by M. A. Arbib. Cambridge, Mass.: MIT Press.

Ramachandran, V. S., and S. Blakeslee. 1999. *Phantoms in the Brain: Probing the Mysteries of the Human Mind*. New York: Quill.

Epilogue

The B-52 bombers in the United States Strategic Air Command have been flying for almost fifty years. While looking similar on the outside, a B-52 of today is nothing like a B-52 of the 1950s on the inside. The avionics systems have been replaced many times over as new technologies have been developed. The engines have been replaced as they have worn out. Internal struts and beams that showed signs of cracking have been replaced during comprehensive overhauls. The metal skin elements have been replaced as necessary over the years or when new materials have been developed. The hydraulic lines and valves have all been replaced during servicing. In short, a B-52 of today with any given airframe number is materially almost entirely different from the B-52 of forty-five years ago with the same airframe number. But it is still the same B-52, just made of different material. And there is no end in sight to the lifetime of that B-52. Its subsystems and structural members will continue to be upgraded and replaced for years to come. B-52s are immortal.

With the developments described in chapter 10, are we destined to become living B-52s, immortal through replacement of all our constituent parts?

Certainly those of us alive today will benefit from the new technologies of silicon and steel becoming integrated into our bodies. Our lives will be extended, perhaps significantly. Already our mobility has been extended in time through artificial joints. The new implant technologies will let us extend the lifetime of our sensors, our eyes and ears. Many of our frailties that come from loss of motor control, through degradation of nerves, or brain disease, will also be repairable. These ways of extending our useful and enjoyable lives are independent of augmenting our capabilities beyond the ordinary younger human capabilities. That will happen in parallel but is irrelevant here. The question is whether we will be able to replace everything while in the silicon-and-steel era.

In chapter 9, I argued that the complexity of our embodiment, the systems-level symbioses, made it unlikely that we will be able to simply download our brains into a computer anytime soon. We must remain mortal for now, and we must face the challenges that mortality imposes upon us. The lives of our grandchildren and great-grandchildren will be as unrecognizable to us as our use of information technology in all its forms would be incomprehensible to someone from the dawn of the twentieth century.

The world is changing, and our humanity within our world is changing with it. The forces of change are irresistible, as they have been for the last five hundred years. Wishing they were not so will do us no more good than such wishing has done for many in past generations.

The future is best approached with an open mind, with an understanding of our deeply held prejudices, and with a willingness to reexamine the nature of our humanity. Questions about it are going to be forced on us. Blind allegiance to deeply held beliefs will serve us no better than ignorant whining has in the past.

Epilogue

Just as scientific revelations have made us understand and appreciate the universe better than we were able to with only scraps of faith to guide us in past explanations, so too will our new revelations, brought on by technological demonstrations and innovations, allow us a deeper understanding of what truly we are. And then we will have the power to change that.

Appendix.   How Genghis Works

Chapter 3 briefly explained some of how Genghis operated. Here we give the complete details of its software, not hiding anything. First, however, we need to examine a few details of Genghis's mechanical construction and sensor suite so that we can understand how the software made it operate.

Genghis had six legs arranged just like those of a six-legged insect—three pairs along the length of the body, as shown in figure 5. Unlike real insects, however, each of Genghis's pairs of legs were the same; real insects often have much larger rear legs and much smaller front legs. Figure 6 shows the kinematic arrangement of Genghis's legs. Each leg had two motors attached to it, and each leg could swing from its "shoulder" back and forth and up and down under the

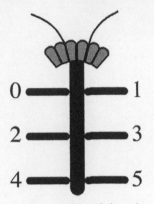

Figure 5. Genghis has six legs, three per side, arranged symmetrically. They are numbered as shown.

control of the two motors. Genghis's legs had no "knees," "ankles," or "toes," so there were only these shoulder motors. We will refer to the motor that swung the leg back and forth as the alpha, or α, motor for *advance*. The motor that swung the leg up and down will be called the beta, or β, motor for *balance*.

Each motor was controlled by giving it a number between -25 and 25 inclusively, representing a position to which it should move. For α motors, 0 made the leg point straight out orthogonal to the body, negative numbers set the position of the leg to point backward, and positive numbers made the leg reach forward. For the β motors, 0 made the leg stick straight out parallel to the floor, while positive numbers made it point downward and negative numbers made it point upward.

When a motor was given a number, corresponding to a commanded position, it accelerated to a maximum velocity to move the leg toward that position, then decelerated to reach zero velocity at the target position. A motor could be made to go slower than its maximum velocity by *leading* it— i.e., by feeding it a series of commanded positions just beyond its current position.

Genghis had four sorts of sensors.

- There were two whiskers up front that were either ON or OFF, depending on whether they were touching something with enough force to bend the whisker.
- There were force sensors on each motor. When a leg was blocked from reaching its target position, the force on the leg (in each of the

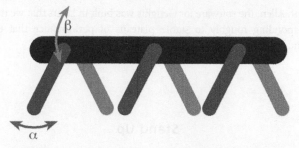

Figure 6. Each of Genghis's legs has two degrees of freedom controlled by separate motors. The α motor swings the leg backward
and forward, while the β motor swings the leg up and down.

α and β directions) could be measured as a number in the range 0 to
15, where larger numbers meant more force.

- There was an inclinometer made from a semicircular trough with a
  ball bearing that could roll back and forth. It measured the angle
  that the robot body was pitched up in the front, giving a value of 15
  at the extreme, or giving a value of 0 in the extreme of being pitched
  up in the back.

- There were six pyroelectric sensors (behind plastic Fresnel lenses)
  arrayed at the front of the robot, each of which registered as ON or
  OFF depending on whether it detected a change in infrared radiation
  levels. These sensors were sensitive to radiation with a wavelength
  of approximately 10 microns, which happens to correspond to the
  temperature of people.

The sensors provided inputs to the computations onboard Genghis, while
the motor commands were the ultimate outputs of the overall computation. Internally, however, we arranged Genghis's computations as a collection of little
tiny computers (which in chapter 3 we called augmented finite-state machines,
or AFSMs) that ran continuously, with virtual wires joining internal outputs of
AFSMs to internal inputs to AFSMs. Figure 2 (p. 47) shows the complete wiring
diagram of Genghis's computations. This diagram represents fifty-one different
AFSMs, many of which are six-way duplicates, one for each leg.[1]

1. In the actual Genghis code there were fifty-seven AFSMs, not fifty-one, as there were
six extra machines used to filter out noise on the downward force sensors—the details
are not important in understanding how Genghis works, as a better sensor would have
had precisely the same effect.

As with Allen, the software for Genghis was built in layers that we thought of as corresponding roughly to stable plateaus of performance that evolution might find.

Stand Up

The simplest behavior for Genghis was for it to stand up. To establish this behavior twelve simple finite-state machines were needed, one for each motor on Genghis. Six identical machines controlled the α motors, and six identical machines controlled the β motors. Here are the definitions of those machines.

> *alpha pos.* Starting with a default position of 0, continuously set the motor to that position.

> *beta pos.* Starting with a default position of 20, continuously set the motor to that position.

Note that the *alpha pos* and *beta pos* machines were identical except that they had a different set point to which they drove their attached motor. Further, note that none of these finite state machines has any input or output, so there were no virtual wires connecting them.

With this set of machines Genghis was able to do precisely one thing. When it was switched on, it stood up—as long as it was lying on its belly on a flat floor. If it happened to be lying on its back when switched on, it simply moved its legs to the stance position, and they ended up pointing skyward, and Genghis did not move at all. This layer of control thus worked only in the right ecological circumstances but produced inappropriate behavior in other circumstances. All animals have such fallibilities, but usually they are less pronounced. On later robots that we built, with better mechanical capabilities, we were able to incorporate behaviors to recover from being upside down, although the simplest layer of behavior was no more complex than that of Genghis.

Involuntarily Step

Our next layer of evolutionary progress was to give Genghis the ability to make a single step forward, with appropriate pushing backward on legs still on the ground. In this layer Genghis had no reason to step forward, but if a person lifted one of its legs, it would.

The first thing to do was to modify the *alpha pos* and *beta pos* machines so that they had an input, and an output. The new definition for the *beta pos* machine follows; the definition for *alpha pos* is just the same, save that it retained its old default position of 0.

> *beta pos.* Input: δ. Starting with a default position of 20, continuously set the motor to that position, and output the most recently commanded motor position. If a new δ is received, change the current target motor position by that much.

This new *beta pos* machine takes commands on its input that tell the motor how to move relative to its current position and to maintain that new position as a consistent goal.

We next built feedback loops for both α and β motors. The six independent feedback loops for the β motors were made by connecting the output of a *beta pos* machine to the input of a new *leg down* machine and connecting its output to the input of the same *beta pos* machine. Figure 2 illustrates this connection. The simplest version of the *leg down* machine just tried to force the motor back to its down-pointing position of 20, no matter how the set position inside *beta pos* may have been altered.

> *leg down.* Input: β. Continuously output 20−β.

The effect of adding such a machine into the network controlling each leg at first seems rather trivial. Its strength will become apparent when we add more network, which occasionally sends a command to a *beta pos* machine to lift a leg. Very quickly the leg will be brought back down to the stand position.

In order to make the leg motions a little smoother, the actual definition used for *leg down* was slightly more complex.

leg down. Input: β. Continuously output max (8 −β, min (20−β, 4)).

The added complexity made for a fast, but not too fast, downward swing, until close to the target position, and then a gradual easing in to that target.

The feedback loop for the α motors was a little more complex. A single *alpha balance* machine participated in a loop with all six *alpha pos* machines.

> *alpha balance.* Six inputs. Continuously add up the six inputs. Whenever the result is bigger than 5 or smaller than −5, divide it by 6, negate it, and output that as the result.

The six inputs to the *alpha balance* machine, as shown in figure 2 (p. 47), reported the current position of each α motor, i.e., the amount of swing forward (positive) or backward (negative) from the straight out position (zero). The *alpha balance* machine had only one output, but it was connected to the δ inputs of each of the six *alpha pos* machines. The *alpha balance* machine tried to get the sum of the outward pointing, α, angles of all six legs to be zero. It was not trying to get all six of them to be zero, just their sum. The *alpha balance* machine became useful when six *alpha advance* machines were added.

> *alpha advance.* Inputs: α, β. If β is less than 5, then output 15−α.

The inputs to each *alpha advance* machine were the reports of the commanded α and β positions, as can be seen in figure 2. When an *alpha advance* machine detected that a leg was up in the air, by the fact that the β position was less than 5, it commanded that same leg to swing forward. But to understand the dynamics, we must see how the *alpha advance* machines were connected to their corresponding *alpha pos* machines.

In the diagram in figure 2 the output of each *alpha advance* machine *suppresses* the connection between the output of the *alpha balance* machine and the δ input to the *alpha pos* machine. When a message is sent along a virtual wire that suppresses another, the signal gets through immediately, but no signals can traverse the original wire for 80 milliseconds.

Thus when the *alpha advance* signal noticed that the leg was up in the air, it repeatedly sent signals to force the leg forward. By forcing the leg forward it made the sum of the inputs in the *alpha balance* machine go positive, so it started sending out commands to all the legs to move backward a little. Except that the leg that had just moved forward did not receive any of those messages—they were suppressed, until the outward pointing angles of the legs were again in balance by summing to zero. But notice that meanwhile the *leg down* machine would have ensured that the leg was brought back down to the stance position, eventually shutting off the stimulus to the *alpha advance* machine, so that it would no longer be sending out commands to swing forward.

The dynamics of the interactions of all these finite-state machines are perhaps a little hard to see, although we have gone through each component of what would happen already. The overall effect is as follows. If any leg was up in the air for any reason, then it swung forward and down. The legs that were still on the ground all moved backward a little, pushing the body of the robot forward, so that when the forward-swinging leg touched the ground it would be in a step-forward position. This is the basis of walking. It is exactly the right coordinated motion of all six legs, arising from very simple local dynamics on each leg.

Voluntarily Walk

The next layer of AFSMs lets the robot walk voluntarily. Six *up leg trigger* machines were added.

> *up leg trigger.* Inputs: τ, β. Whenever an input is received on τ, for the next half a second repeatedly output $4-\beta$.

Each one had a triggering input τ, and also an input that reported the current commanded position of the β motor. Whenever a trigger input was received, an *up leg trigger* machine forced the leg, this time via *suppression* of the basic β feedback loop, to lift up to position 4 and to stay there for half a second. This was sufficient to trigger the involuntarily step response we have already discussed, so the robot would take one step forward.

To make Genghis walk, it was now necessary to add but one final AFSM.

walk. Six outputs: τ_0, τ_1, τ_2, τ_3, τ_4, τ_5. Every four-tenths of a second send something to an output in the following repeated order: τ_4, τ_1, τ_2, τ_5, τ_0, τ_3.

The outputs of the *walk* machine were each connected to the τ input of the corresponding *up leg trigger* machine. Thus the *walk* machine repeatedly triggered each leg to lift up and induce the involuntary step response. The way we have specified it here, Genghis walked with what is known as a *ripple gait*. On each side of its body first the back leg would step, then the middle leg, then the front leg. The two sides were exactly out of phase so that the back leg on the right side would step exactly halfway between two steppings of the back leg on the left side. Notice that with the timings specified for both the *walk* machine and all the *up leg trigger* machines, more than one leg was swinging forward at all times. With a slightly different *walk* machine that lifted legs 0, 3, and 4 simultaneously, and then legs 1, 2, and 5 simultaneously, Genghis was able to walk faster with the *alternating tripod* gait.

With these layers of control Genghis was able to walk well over flat ground, but it did not really appear to be lifelike. Rather, it appeared to be a clockwork-like machine, as it did not take into account any obstacles in its way, or unevenness of ground, and it did not seem to have any purpose.

Walk over Obstacles

To make Genghis handle uneven terrain, we made use of the force sensors in each leg. The most obvious thing to do was to make it lift its legs higher whenever it encountered an obstacle so that it could walk over it. This required a small modification to the *up leg trigger* machines.

> *up leg trigger.* Inputs: τ, β, υ. Start with a default υ of 4. Whenever an input is received on τ, for the next half a second repeatedly output $\upsilon - \beta$, and then reset υ to be 4 again.

The new υ input provided a new temporary height to which the leg should be lifted. It was supplied by six new AFSMs, the *alpha collide* machines.

> *alpha collide.* Inputs: α, β. If α is 15 and β is bigger
> than 7, and if the force felt on the α motor stays at 15
> for 80 milliseconds, then output -10.

These machines had inputs reporting the most recently commanded positions of the two leg motors. If the leg had been commanded to go all the way forward in a step, and if the leg had started on its downward trajectory, and if the α motor felt a large consistent force, then the *alpha collide* machine reset the υ input on the appropriate *up leg trigger* to force the leg to lift higher on the next step. With these machines in place Genghis was able to fairly gracefully climb over obstacles in its path, once it had run into them. The colliding leg would swing up higher and step over any obstacles the height of Genghis.

Genghis was now responding to obstacles in its way, but it did not do a very good job of balancing laterally, side to side. If, say, its middle right leg was on a rock, then its whole body would tilt to the left, without its other two right legs even touching the ground. To fix this behavior we added six more AFSMs, the *beta balance* machines.

> *beta balance.* If the force felt in the β motor is 7 or
> less, then do nothing; if it is 11 or more, then output
> -3; otherwise output 0.

Each *beta balance* machine ensured that there was not too much weight on a single leg, such as when one leg was on a high obstacle. The output of the *beta balance* machine was piped through to the lower level β feedback loop, and it set a new set point, not pushing the leg down any farther, or even lifting it up a little in high-force situations so that the robot became more balanced. As can be seen in figure 2, the output of a *beta balance* machine merged with the output of an *up leg trigger* machine. The *beta balance* machine provided the *default* signal on the virtual wire that then suppressed the β feedback loop. Whenever the *up leg trigger* machine provided an output, it overrode what the *beta balance* machine was trying to do, as the *up leg trigger* was trying to raise the leg for stepping forward. On the other hand, it was important that when the leg was down and a high force was being felt, that the *beta balance* machine continued to output signals, even if they were commands to change the leg position by 0. That overrode the natural tendency of the β feedback loop to push the leg down to its default position.

With these additions Genghis was able to respond to its environment and clamber over fairly rough terrain. It sometimes scraped its belly, and missed many footholds, but it had persistence and was able to eventually push its way through most situations in which we placed it.

Walk Competently

Genghis was still not a very elegant walker. Most noticeably there was a bad side effect of the *beta balance* machines. As the robot started to climb an incline, more weight was transferred to the rear legs. The force felt by those legs was interpreted by the attached *beta balance* machines to mean that the legs were pushing down on some sort of obstacle, so their set position was raised and the robot sat down on its rear legs. Likewise, when going down an incline the robot would tend to nose-dive into the slope.

Both these problems were alleviated by the evolutionary addition of two *pitch* AFSMs, one for the back pair of legs and one for the front pair of legs.

> *pitch.* If the pitch sensor is within 5 units of the appropriate extreme output an inhibitory message.

Each *pitch* machine sent its output to inhibit the outputs of the *beta balance* outputs of each of its pair of legs. The rear *pitch* machine would output a message when the pitch sensor was saying that the robot was close to its maximum nose up pitch. The front *pitch* machine made a corresponding output when the nose was pitched down. These outputs were connected to the outputs of their appropriate *beta balance* machines to inhibit any signal on those virtual wires for 80 milliseconds whenever an inhibitory message was sent. The effect was to stop the robot from falling on its rear end when climbing, and to stop it from nose-diving when descending, a marked improvement in walk elegance.

The walking style of Genghis was also significantly improved by linking each of the front whiskers to the appropriate front leg via one of two *whisker* AFSMs.

> *whisker.* If the whisker on this side is ON, then output −10.

The outputs of the two *whisker* machines were connected so as to force the appropriate front leg, via the υ input to the *up leg trigger* machine, to lift up higher on the next step. In this way, as Genghis walked forward, if its whiskers encountered an obstacle, its legs would lift high over the top of it, without having to wait to encounter high forces by running into it.

With these additions Genghis walked quite competently. Alas, it appeared to have no purpose in life and so was not quite creaturelike in its behavior.

Chase Prey

We already examined the last layer of Genghis's control software in chapter 3. That last layer turned Genghis into a predator. We flesh out the details more fully here.

A single AFSM, *IR sensors*, was added to summarize what the pyroelectric sensors had seen in the previous half second. Its output was connected to another single machine, named *prowl*, with one output that simultaneously inhibited all six ouptuts of the *walk* machine.

IR sensors. Continuously output a list of which of the six pyroelectric sensors have triggered ON within the previous half a second.

prowl. Input: pyroelectric sensor list. Continuously output an inhibitory message, but if ever any sensors are listed on the input, then stop doing so for 5 seconds.

The combined effect of these two machines was to make Genghis not move when it saw no infrared activity in view of its pyroelectric sensors. The *prowl* machine inhibited all the outputs of the *walk* machine so there were never any triggers to *up leg trigger* machines to lift any leg and initiate a step of the robot.

When there was infrared activity in the sight of the forward-looking pyroelectric sensors, then the robot started to walk for a few seconds and would continue to do so if it maintained a view of infrared activities.

Following this mindless physical coupling, we were able to get Genghis to track and follow an infrared source. First, we had to add one more input to the *alpha pos* machines.

alpha pos. Inputs: δ, μ. Starting with a default position of 0, continuously set the motor to that position and output the most recently commanded motor position. If a new δ is received, change the current target motor position by that much, but never go below the most recently received μ, and always reset μ to −25.

The μ input provided a backstop beyond which the legs could not swing. By changing μ to all the legs on the right side, say, to make those backward pushes smaller, the robot ended up taking smaller steps on the right side, while maintaining its normal-sized steps on the left side. The robot thus turned to the right. A final single AFSM, named *steer,* was added to capitalize on this capability.

> *steer.* Input: pyroelectric sensor list. Continuously count the number of left sensors and right sensors that are ON. If there are more on the left, then output 7 on the left side, and if there are more on the right, output 7 on the right side.

The left-side output went to the μ inputs on each of the left-leg *alpha pos* machines, while the right-side output when to the right-leg machines.

Now Genghis would sit and wait for some moving source of infrared to come along and pass in front of it. Then Genghis would start walking toward it, steering left and right to keep the target in sight, scrambling over any obstacles in its way.

Index

Index

artificial life (Alife) *(continued)*
  failure to evolve intelligent creatures,
    possible reasons for, 184–87
  juice hypothesis and, 187–91
Asimo robots, 71
Asimov, Isaac, 71n, 72
Asimov's laws, 72–73, 199, 203–4
astronomical discoveries, 160–62
"As We May Think" (Bush), 9–10
Attila (robot), 54–55
auditory systems for robots, 94–95
augmented finite-state machines (AFSMs),
    46–51, 53, 54–55
automobiles:
  automatically driven, 26–27
  manufacture of, 114–15
  navigation systems, 130

Bacon, Roger, 13
Bajscy, Ruzena, 43
Ballard, Dana, 81–82, 83
Ballistic Missile Defense Organization
    (BMDO), 58–61
bandwidths, 134–35
batteries for robots, 129
Baumgart, Bruce, 26–27
beacon systems, 116
beer-fetching robots, 125–26
Behavior Language, 54
beingness, 50–51, 149–50
  in animals, 51, 150–54
  emotions of humanoid robots and, 156–59
  see also consciousness in robots
Bekey, George, 43
belief systems, 159–60, 161
Berners-Lee, Tim, 10
Biak Islanders, 68
Bicentennial Man, The (film), 73, 199
biomolecular nature of human beings,
    172–74
biotechnology, 11, 232–36
Bit (robot), 108–9
Bodicea (robot), 55
body augmentation: *see* human-machine
    hybridization
Brady, Michael, 42
Brahe, Tycho, 161
brain death, 231
Breazeal, Cynthia, 54–55, 64, 68, 91–92,
    93, 95–96
Brooks, Rodney A., 6, 33
  computation organization for robot
    control, 35–44

early interest in robotics and artificial
    intelligence, 27–28
  My Real Baby robotic dolls, 107–13
  research heuristic of, 37, 40
  software organization providing lifelike
    behavior in robots, 46–51
  space exploration with robots, 55–61
Bullock, Bruce, 57–58
Bush, Vannevar, 9–10

cargo cult science, 68
Carroll, Lewis, 18
Cart (robot), 25, 26–27, 28–31
category theory, 190n
Catholic Church, 161
cats, 153
cave paintings, 12
cellular telephones, 227
Chalmers, David, 176, 177–78, 187, 194
Chatila, Raja, 24
chess playing by machines, 165, 168–69,
    186
chimpanzees, 3–5, 151, 152–53, 164
"Chinese Room" thought experiment,
    179–80
Chomsky, Noam, 165
civilization revolution, 7, 9
Clark, Jim, 84
Clarke, Arthur C., 198, 214
cleaning robots, 115–26
  almost life-forms, 121
  beacon systems for, 116
  clothes-cleaning, 123, 124
  convenience of, 121
  cost concerns, 122
  coverage of floor area to be cleaned,
    116–18, 120–21
  discreet approach to cleaning, 122
  ecology of robots living and working
    together, 118–20
  emergent set of interacting behaviors, 120
  for frivolous sorts of tasks, 123, 125–26
  odometry and, 117
  pucksters, 119–20
  recharging capability, 118
  remote-presence robots as, 146–47
  robots currently on the market, 120–21
  table-setting and -clearing, 124–25
cochlea implants, x, 215–16, 217
Cog (robot), 68–69, 87–88, 91, 149
color agnosia, 192–93
color perception, 76
Colossus: The Forbin Project (film), 199

Index

computation:
   juice hypothesis and, 188–91
   organization for robot control, 35–44
computers:
   first digital computers, 15–16
   increases in computational power, 197
   integrated circuits, 21–22
   Lisp Machines, 35
   wearable computers, 214
   see also artificial intelligence; robotics
computer vision systems, 37–38
   achievements and shortcomings of
     current technology, 90–91
   attention systems, 93–94
   calibration procedure, 149–50
   difficulty of computer vision, 73–74
   juice hypothesis and, 190
   mechanical aspects of human visual
     system, duplication of, 80, 88
   processing of images, 88–89
   speed requirements, 87–88
   surgical uses, 223–24
   three-dimensional models, 28–29, 33
   usual and unusual movements, recogni-
     tion of, 140–41
conferences using remote-presence robots,
   142–43
Connell, Jonathan, 43–44, 52
consciousness downloaded into silicon,
   205–6, 207, 208
consciousness in robots, 50–51
   arguments against potential for, 176–80
   as beyond human capability to create,
     191–93
   failure of robots to achieve consciousness,
     possible reasons for, 184–87
   juice hypothesis and, 187–91
   potential for, 174–76, 180
   slavery issue, 195
   uncertainty regarding nature of con-
     sciousness and, 194–95
   visceral responses, 157
Constable, Bob, 186
Copernicus, 160
Crick, Francis, 164
cryogenics, 206–8
Cyborgs, 214; *see also* human-machine
   hybridization

Damasio, Antonio, 156–57, 180
Darby, Abraham, 8
Darwin, Charles, 162, 164–65
death, fear of, 208

Deep Blue computer program, 169 and *n*,
   170, 178*n*, 186
Dennett, Daniel, 17
Desai, Rajiv, 57
digger wasps, 82–83
digital cameras, 75–76
Digital Creature Laboratory, 105
digital revolution, 5–6
disruptive technologies, 100–101
DNA, 164
Dr. Strangelove (film), 203
dogs, 148, 150, 153, 216
   robotic dogs, 17, 105–7
Doi, T., 105
Dreyfus, Hubert, 168–69

early adopters, 106
Earnest, Lester, 25, 26
E. coli robots, 234–35
electroencephalography, 17
electromagnetic relays, 15
electronic creatures, 15–21
Electrotechnical Laboratories, 71
ELIZA computer program, 166–67
embodied robots, 51–55
emergent behaviors, 19–20, 21
Engelberger, Joe, 114
ENIAC computer, 16
evolution:
   artificial life and, 181–84
   human response to concept of, 162–63
   robot control systems and, 40
extropianism, 208
eyes: *see* human visual system

Faruqee, Hamid, 136
"Fast, Cheap, and Out of Control: A Robot
   Invasion of the Solar System" (Brooks
   and Flynn), 56
"fast and cheap" robots for space exploration, 55–62
fear, 150–51
fish, 151, 153
Flynn, Anita, 33, 34, 41, 55, 56
Frankenstein (Shelley), 198
friendly robots, 137
Fujitsu Research Labs, 71
Furby toys, 104–5

Galileo, 161, 164–65
genetic engineering, 234–36
genetic modification, 234
genetics, specialness of humans and, 164

Index

Index

human visual system, 74–87
blind spots, 77
changes in world, failure to detect, 83–84
color perception, 76
as complex arrangement of partial solutions to difficult problems, 84
eye components, 76–77
eye modification technologies, 217–18, 225–27
eye movements, 78–79, 86
gaze direction, 85–87
interpretation of looking at world, 77–78
mechanical aspects, 80
processing of images, 88–89
saccades, 78–79, 82
search for and storage of information relevant to a task, 81–83
smooth pursuit, 79
social interaction and, 85–87
stereo vision, 79–80
vestibular-ocular reflex, 79
hypertext systems, 9–10

I, Robot (Asimov), 72
identity chips, 216
immortality through technology, 204–8
industrial revolution, 8–9
industrial robot arms, 37, 114–15
information revolution, 6, 9–10
insects, 40, 45, 51, 152, 153–54
integrated circuits, 21–22
intelligence:
  characteristics of, 36–37
  see also artificial intelligence
intelligent entities, 103–7, 112–13
intelligent robots of the future, 197–209
  control issue, 202–4
  damnation scenario, 198–204
  empathy for humans, 201–2
  energy self-sufficiency, 202
  inevitability of, 209
  reproduction by, 200–201
  salvation scenario, 204–8
intentionality: *see* beingness
International Symposium of Robotics Research (1985), 42
Internet, 10, 73n
  direct mental access to, 228, 229, 230
  see also remote-presence robots; World Wide Web
Internet-ready appliances, 139
Intuitive Surgical company, 224
Irie, Robert, 91

iRobot Corporation, 34n, 45n, 58, 107, 111–12
iRobot-LE (robot), 131–32, 133, 134
  photo of, 131
ironing, 124
IT (robot), 108

Jacquet-Droz, Pierre and Henri-Louis, 15
Japan:
  humanoid robot development, 69–71
  labor shortage, 135–37
  organ transplantation, 231, 232
Jet Propulsion Laboratory (JPL), 24, 56–57
Jobs, Steve, 26
Johns Hopkins University, 202n
Johnson, Mark, 67
juice hypothesis, 181, 187–91, 198
Juro, Wada, 231

Kasparov, Garry, 169 and n, 170, 186
Kato, Hirokazu, 69
Kawasaki Heavy Industries, 114
Kelly, Kevin, 57n
Kepler, Johannes, 161
Ketterer, Anton, 14
kinetic kill vehicles, 58–59, 60
Kismet (robot), 64–65, 92–97, 148–50, 156
  photo of, 93
Knight, Tom, 234–35
Kubrick, Stanley, 74, 198
Kurzweil, Ray, 205, 206

Laboratoire d'Analyse et d'Architecture des Systèmes (LAAS), 24
labor shortages alleviated with remote-presence robots, 135–37
Lakoff, George, 67
language acquisition by robots, 96n
language and thought as based on metaphors for bodily interactions with world, 67–68
Lanier, Jaron, 167–68, 206
Larson, Noble, 35
lasers, 122–23
lawn-mowing robots, 127–30, 134
learning capabilities, 20, 21
legs, robotic prosthetic, 214–15, 218
Leonardo da Vinci, 14
Life on the Screen (Turkle), 149
Lipson, Hod, 184
Lisp Machines, 35
Lozano-Perez, Tomas, 37
Lynch, Jim, 157, 158

Index

MacHack computer program, 169
Machina docilis, 19–20
Machina speculatrix, 17, 18
macular degeneration, 217
Madhani, Akhil, 224
Maes, Pattie, 206
Maillerdet, Henri, 15
Mann, Steve, 214
Marr, David, 38
Mars, robot exploration of, 61–62
Mataric, Maja, 53
mathematics, 165, 169–70
    juice hypothesis and, 188–91
Matrix, The (film), 199
McCarthy, John, 25, 26, 35
McMann, Tom, 214–15
mechanical creatures, 13–15
mental-tapping technology, 227–30
metaphors for higher-level concepts derived
    from bodily experiences, 67–68
mice, 150–51, 153
Miller, David, 57
mind-body dualism, 172
Minsky, Marvin, 25, 74, 204
MIT Artificial Intelligence Laboratory, 25,
    35, 213–14
Mitchell, Bill, 10
MIT Laboratory for Computer Science,
    213–14
MIT Media Laboratory, 214
Moon, robot exploration of, 24–25
Moore's law, 129, 185–86, 205
Moravec, Hans, 27, 34
    Cart project, 28–31
    salvation scenario, 204, 205, 206
    three-dimensional reconstruction
        program, 28–29n
More, Grinnell, 34, 45
Morris, Errol, 57n
Moses, Joel, 170
movies' portrayal of robots, 115, 198,
    199–200
M2 (robot), 71
Muhleisen, Martin, 136
music distribution, 100–101
My Real Baby robotic dolls, 107–13, 156,
    157–58

Napster, 100
Narayanan, Sathya, 33–34, 41
NASA, 24, 57, 61–62
navigation by robots, 129–30
Neanderthals, 5

Nelson, Ted, 10, 100
Nerd Herd robots, 55
neural networks, 103
Newcomen, Thomas, 8
Newell, Allen, 35
"new stuff":
    juice hypothesis, 181, 187–91, 198
    rejection of "man as machine" perspective
        and, 176, 177–78
Nicolelis, Miguel, 220–21
night vision enhancement, 226
Nilsson, Nils, 22
Ning, Peter, 33, 35, 41
Norman (robot), 27

odometry, 117
On the Origin of Species (Darwin), 162
orangutans, 152
O'Regan, Kevin, 84
organ transplantation, 231, 232

paintings and sculptures as artificial
    creatures, 12–13
Papert, Seymour, 74
Parkinson's disease, 222–23
Pavlov, Ivan, 20
Penrose, Roger, 176–77, 191–92, 194
Pentland, Sandy, 37–38
Pepperberg, Irene, 4
photographic memory hypothesis, 81–83
physics, 163–64
pixels, 37–38
Pollack, Jordan, 184
Polly (robot), 53–54
Power, Donald C., 25n
Pratt, Gill, 71, 72, 215, 218
predicting the future, 99–101
Prescott, Tony, 51n
Principia Mathematica (Russell and White-
    head), 165
printing technology, 9
prosody, 94–95
P series robots, 70–71
psychotherapy, 166–67
puckster robots, 119–20
PUMA robots, 114

quantum mechanics, 163–64, 177

Raibert, Marc, 44
rats, 153
Ray, Tom, 181–82
Reagan, Ronald, 58

258

Index

relativity, theory of, 163–64
religion, 159–60, 161, 162, 163, 174, 208
remote-presence robots, 131–47, 221
  bandwidth needed for, 134–35
  conferences using, 142–43
  control mechanisms, 133–34
  demonstrations of, 132
  dexterous operations, 145–46
  door- and lock-operating capabilities, 143–45
  housework performed by people in poorer countries via, 146–47
  labor shortages alleviated with, 135–37
  lag between command and action, 133
  mental-tapping technology and, 230
  monitoring of vulnerable persons with, 141
  security functions, 139–40
  simple domestic uses, 138–39
  special interest groups' use of, 142
  spying with, 140–41
  surgical uses, 224
Rensink, Ron, 84
Repin, Ilya, 81
reptiles, 151, 153
retinal implants, x, 217–18, 225–26
retinas, 77
Robbie (robot), 56
Roberts, Larry, 73–74
robotics:
  batteries for robots, 129
  biological matter incorporated into machines, 233–34
  computation organization for robot control, 35–44
  early digital robots, 22–31
  electronic robots, 17–21, 23
  embodiment principle, 51–55
  friendly robots, 137
  genetically engineered robots, 234–35
  industrial robot arms, 37, 114–15
  lawn-mowing robots, 127–30, 134
  movies' portrayal of robots, 115, 198, 199–200
  navigation capabilities, 129–30
  simulator testing of robot capabilities, 41–42
  situatedness principle, 51–55
  software organization providing lifelike behavior, 46–51
  space exploration with robots, 24–25, 55–61
  surgical aids, 224–25

toys, robotic-based, 104–13
  walking capability, 44–51, 54–55
  see also cleaning robots; consciousness in robots; human-machine hybridization; humanoid robots; intelligent robots of the future; remote-presence robots
robotics revolution, 10–11
Rocky robots, 57, 61
Rosen, Robert, 190*n*
Rosenberg, Chuck, 107, 108
Ross, Thomas, 17
Russell, Bertrand, 165

SAIL computer, 25–26
Salisbury, Ken, 224
Scassellati, Brian, 93
Scheinman, Victor, 114
Schmidt, Rodney, 26
Scott, David, 32–33, 58, 59
search engines, 102
Searle, John, 176, 178–80, 194
security for the home, 139–40
Shakey (robot), 22–23, 38
Shannon, Claude, 17, 168
Shelley, Mary, 198
Sheridan, Tom, 133
Shirley, Donna, 57
Simon, Herbert, 44
Sims, Karl, 182–84
situated robots, 51–55
Slagle, Jim, 170
sleep, 155
social interaction:
  with humanoid robots, 68, 87, 92–97, 148–49
  human visual system and, 85–87
software organization providing lifelike behavior in robots, 46–51
  Genghis's software details, 244–52
Sojourner (robot), 61
sonar, 129–30
Sony Corporation, 71, 105–7
Sozzie (robot), 117–18
space exploration:
  "fast and cheap" robots for, 55–61
  lunar unmanned landers, 24–25
spying with remote-presence robots, 140–41
Stanford Artificial Intelligence Laboratory (SAIL), 25–26, 114
Stanford Research Institute, 22
Starner, Thad, 214

259

Index

Star Wars films, 115
Stauffer, Chris, 140–41
steam engines, 8
stimulus-response mechanism, 151
StockMaster.com, 135n
subsumption architecture, 35–44
Sun Microsystems, 26
supersapiens, 193
surgery, x, 223–25
Surveyors (unmanned space vehicles), 24
Sussman, Gerry, 74
symbolic mathematics done by machines,
    169–70
syntax, 3–4, 5, 165

Takanishi, Atsuo, 70, 71
Tamagotchi toys, 104, 110–11
Tau, Robert, 134–35n
technological revolutions, 5–11
technology, 3, 4–5
telecommuting, 10
telegraph, 9
telephone, 9, 15
television, 75–76
Teller, Edward, 58
Terminator, The (film), 199
3G telephones, 139
Tierra computer program, 181–82
Tiger Electronics, 104
Tokyo University, 71
tool use, 4–5, 6–7
Tooth (robot), 57
Torrance, Mark, 134–35n
tortoises (robots), 17–21, 23, 30
Toto (robot), 53
toys, robotic-based, 104–7
    My Real Baby robotic dolls, 107–13, 156,
        157–58
transhumanism, 208
transistors, 21
translation of languages, 101–2
Turing, Alan, 166, 168
Turing machines, 177
Turing test, 166–67
Turkle, Sherry, 149
2001: A Space Odyssey (film), 63–64, 74, 84,
    97, 198

Unexpected Return (Repin), 81
Unimate (robot), 114
University of the West of England, 202
urban robots, 34n

vacuum tubes, 15–16
Vaucanson, Jacques de, 14
VECTROBOT (robot), 34
Velásquez, Juan, 105–6
video games, 103–4
vision systems: see computer vision systems;
    human visual system

Wabot robots, 69–70
walking capability:
    in humanoid robots, 70, 71
    in six-legged robots, 44–51, 54–55
Wallace, R. A., 17
Walter, W. Grey, 17–21, 23, 30
Waseda University, 69, 71
water clocks, 13
Watson, James, 164
Watt, James, 8
Web servers, 134–35n
Weiss, Ron, 234–35
Weizenbaum, Joseph, 166–68
wheelchairs with robot control systems,
    219–20
Whitehead, Alfred North, 165
Wiener, Norbert, 168
Wilkenfeld, Ari, 218
Williams, Robin, 73
Winston, Patrick, 191
Won, Chi, 108
Wood, Lowell, 58
Woodward, Todd, 192
word processors, 103
World Wide Web, 9–10, 100
    search engines, 102
    Web servers, 134–35n
worms, 152

Xanadu project, 10

Yanco, Holly, 219–20
Yarbus, Alfred, 81
Y2K bugs, 211